Vt gaudet flamma se perimente CVLEX

II

EXPLORATIONS
INTO LIFE ON
EARTH

11 EXPLORATIONS INTO LIFE ON EARTH

CHRISTMAS LECTURES FROM THE ROYAL INSTITUTION

HELEN SCALES

FOREWORD BY SIR DAVID ATTENBOROUGH

Michael O'Mara Books Limited

First published in Great Britain in 2017
by Michael O'Mara Books Limited
9 Lion Yard
Tremadoc Road
London SW4 7NQ

A CIP catalogue record for this book is available from the
British Library.

Papers used by Michael O'Mara Books Limited are natural, recyclable
products made from wood grown in sustainable forests. The
manufacturing processes conform to the environmental regulations of
the country of origin.

ISBN: 978-1-78243-840-3 in hardback print format
ISBN: 978-1-78243-841-0 in ebook format

1 2 3 4 5 6 7 8 9 10

www.mombooks.com
Follow us on Twitter @OMaraBooks

Typeset by Ed Pickford

Cover design: Anna Morrison

Printed and bound by CPI Group (UK) Ltd, Croydon, CR0 4YY

CONTENTS

FOREWORD

by Sir David Attenborough

I have only once in my broadcasting career asked if I could be released from a commitment to make a television programme. That was over forty years ago. I had, a few years earlier, as Controller of BBC2, arranged for the Royal Institution's Christmas Lectures to be broadcast on that network. But now I had resigned from that job to resume making programmes and had recklessly agreed to give one of these annual lecture series myself.

When commissioning the broadcasts, I had stipulated that they were to be transmitted live and unedited, exactly as they were performed. It would add to the fizz and excitement, I thought, if the television audience knew that the experiments really were experiments and that no one could be totally certain of what was going to happen. Now, about to give a series myself, I had to abide by that rule. To make matters worse, I had selected a zoological subject and said I would illustrate it with demonstrations involving wild and unpredictable animals.

I started work on the first programme and it very soon became clear to me that the task was impossible. How could you be sure that the animals would behave in the way that theory dictated? How could I be certain that they wouldn't bite me? Or escape? I laboured away at my script for the first Lecture, trying to get my explanations clear and my experiments accident-proof, at the same time haunted by the thought that I only had the vaguest idea of what the second would be about and no idea whatsoever about the content of the remaining four, apart from their titles. And the series was due to start in a few weeks time.

So I telephoned the BBC producer to tell him that I wished to be released from my contract. 'Nonsense,' he said. I seem to recall that I then offered to make a penalty payment for breaking my contract. He rejected that in similar terms, as I would have expected any producer worth his salt to do when dealing with a nervous contributor. You will get some idea of the result in the seventh chapter of this book.

The Lectures are no longer shown live, as they happen. Perhaps that is just as well from the Lecturer's point of view. But it also suits the television programme planners to be able to show the programmes at a time that suits their schedules and not, of necessity, when they are actually delivered. Nonetheless, I dare say you will marvel at the extraordinary skill and ingenuity with which all my co-contributors devised such dazzling, entertaining and informative experiments and stimulate such interesting thoughts in the minds of their audience. As indeed, I do. But only I – and they – know the full extent of what it can cost to give them.

INTRODUCTION

A few steps away from the busy streets of central London there's a room that's familiar to millions of people around the world; the steep rows of seats, the polished woodblock floor, the large desk. Each year for nearly two hundred years, a scientist has walked into this room to enthuse and enlighten the young audiences who flock to the Lecture Theatre for the Royal Institution (RI) Christmas Lectures. People have watched them on television since 1966 and now Lectures past and present are available to watch online. And after the London Lectures are finished, many of the Lecturers go on tour to entertain audiences from South America to Asia. The founder of the series, the influential British scientist Michael Faraday, would no doubt be astounded to know the Lectures continue today and reach so many people worldwide.

This is the second book celebrating the RI's Christmas Lectures. The first, *13 Journeys Through Space and Time*, took us on a voyage of astronomical discovery and gazed into the solar system and beyond. Now we set off to explore the living wonders of our own planet.

A succession of eminent speakers have filled the Lecture Theatre with a teeming array of furry mammals, luxuriant plants, squawking birds, crawling insects and plenty more besides, as they unlock many of the greatest secrets of life on earth. The book begins in the early twentieth century, a time when studies of the living world were gradually shifting away from a descriptive discipline (preoccupied mainly with finding and naming species) towards the modern science of ecology. Instead of studying organisms in isolation, scientists began to consider the ways living creatures interact with each other and their environment, forming an intricate web of life.

Michael Faraday himself might have been surprised by the topics covered here. Since their inception in 1825 and throughout the nineteenth century, few of the RI Christmas Lectures have focused on the living world. There was one Lecture in 1831 by the Scottish naturalist James Rennie simply entitled 'Zoology' and another, 'Botany', by John Lindley in 1833 (but there is very little archive material to provide details of either). Otherwise, the Lectures tended to concentrate on the physical sciences. Even Charles Darwin's groundbreaking 1859 theory of evolution by natural selection – explaining how life on earth came to be – didn't feature in any nineteenth-century Lectures. Richard Dawkins first tackled evolution in the Christmas Lectures in 1991 (see Chapter 8); perhaps the topic was deemed unsuitable and too controversial for young Victorian audiences.

The Lectures covered in this book represent something of a new generation of Christmas Lecturers. They reflect the growing interest over the last century in understanding how the living world works, together with mounting concerns that

3

human actions are damaging the living systems on which we all depend in so many ways.

Each chapter focuses on a particular Lecture series, which consisted of between three and six hourlong talks originally delivered on several days over the festive period. The aim of the book is to give a flavour of the most exciting discoveries and ideas discussed by each Lecturer and to transport readers on an exploration into the wilds. We'll see how scientists have solved many remarkable mysteries of the living world and revealed ever-greater wonders, from tropical rainforests and the coldest place on earth, to the familiar animals and plants we see in the world around us every day.

CHAPTER I

THE CHILDHOOD OF ANIMALS

Sir Peter Chalmers Mitchell

1911

What better place to start our explorations of life on earth than at the beginning of life for the millions of animal species that live here. Childhood can progress in many ways, as we'll see, from miniature-adults that simply grow bigger, to youngsters that undergo radical transformations. We tour this early world in the company of a zoological expert who has helped rear many young animals. Chalmers Mitchell will show us how tough those early years can be and how young animals do all they can to stay alive.

'Complicated pieces of machinery, like watches or motor cars, resemble animals in many ways, and like them may be new or old, but are never young,' begins Chalmers Mitchell. 'Youth is a property of the living world.' He chooses not to strictly define when childhood begins and ends – the living world is too variable for that – but he does bring into the Lecture Theatre several babies to show the audience what young animals can look like; there's a year-old jaguar, a squirrel monkey, several snakes and a young alligator.

Animals, he says, can be placed in three groups according to their childhoods. The first contains creatures with no period of youth, things like single-celled amoeba that make more of themselves by simply splitting into two identical copies. The second group is where humans belong, together with all the other animals in which the young more or less resemble their parents.

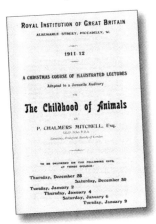

Lecture programme (front cover).

6

Curiously, the young of different species within this second group often look a lot like each other. A reporter from the *Aberdeen Daily Journal* wrote on 29 December 1911 that 'ripples of laughter greeted my ears as I entered the lecture room', as Chalmers Mitchell pointed out how baby gorillas and baby humans look alike, showing pictures on the lantern slide to prove his point. Chalmers Mitchell tells another story of similar babies from the time he imported a young hippopotamus from Africa to London Zoo. A customs officer detained the hippo, convinced it needed examining for infectious livestock diseases because, in his view, it was quite obviously a pig.

In the third group of animals, the young differ so much from the adults that it is almost impossible to work out what they're going to turn into when they grow up. Chalmers Mitchell paints a vivid picture of what it might be like if humans underwent such radical change. He asks the audience to imagine a human baby starting out as a fish, swimming in an aquarium and eating water fleas. When his skin grows too tight, it cracks and splits and a hedgehog creeps out onto land. After eating earthworms in the garden for some time, he once again becomes too big for his skin and, after a second split, climbs out as a fully grown boy.

There are no mammals in this third group, but mainly insects and marine invertebrates, including crabs, lobsters and shrimp, which grow up as a series of distinct larvae. To metamorphose between each stage, they moult their hard exoskeleton and reveal a new, bigger one underneath, which can look radically different to the last, growing new legs, swimming appendages and spines (we'll hear more about insect metamorphosis in Francis Balfour-Browne's Christmas Lectures of 1924–5; see page 37).

Considering the duration of youth, Chalmers Mitchell describes how this can vary enormously between different animals. Elephants are some of the largest and longest-lived animals and also have some of the longest childhoods. Twenty years before Chalmers Mitchell became its secretary in 1921, London Zoo was home to Jumbo the elephant who, at 3.35 m (11 ft) tall, was the world's largest captive elephant at the time. Like all African elephants, Jumbo would have reached maturity only after twenty to twenty-four years. Large size, however, doesn't necessarily tell us that an animal

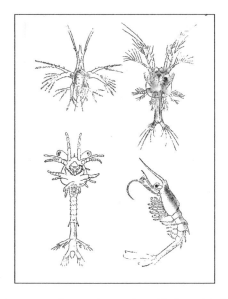

Various stages of prawn larvae, from Chalmers Mitchell's book accompanying his Lectures.

has a long childhood; a hippopotamus becomes fully adult after just five or six years.

Among the insects, many species spend only a very short time as adults. Mayflies live as aquatic larvae in ponds for two years before emerging as adults for just a few hours. Even more extreme are cicadas, which spend seventeen years as larvae buried underground before metamorphosing into adults and flying around in huge numbers; within just two months all the adult cicadas mate, lay their eggs and die.

Some animals never become adults. Chalmers Mitchell holds up a glass container with an animal inside that a reporter at the *Dundee Courier* describes on 2 January 1912 as 'Peter Pan . . . a reptile that never, or hardly ever grows up'. The creature inside looks like a giant tadpole or a newt, with legs and feathery gills sticking out on each side of its head. It is an axolotl, in fact an amphibian, not a reptile, that comes from lakes around Mexico City (in the wild they're now critically endangered because of urbanization and water pollution). Chalmers Mitchell describes how for a long time scientists assumed these unusual salamanders spend their whole life in this state of perpetual youth; they reach sexual maturity without metamorphosing into the adult stage. However, he reveals that some axolotls at the Jardin des Plantes in Paris, kept in tanks with only a small amount of water, surprised everyone when they cast off their gills and developed lungs, as most adult salamanders do. The *Dundee Courier* reporter notes a faint hope among the youngest members of the audience

Metamorphosis of an axolotl from the young, aquatic stage with feathery gills (top), to the land-dwelling, air-breathing adult (bottom).

'that the transformation might take place during the afternoon', but the axolotl didn't proceed to metamorphose in the RI Lecture Theatre.

As all parents know, providing food is a vital part of any young animal's upbringing and Chalmers Mitchell explores this topic from various angles. 'It is a curious fact,' he says, 'that nearly all young birds are fed upon insects, worms and maggots by their parents, even though the adult birds are naturally herbivorous.' Ostriches are one of the few bird species that begin life eating roots, leaves and seeds. The shift in most birds' diet, he points out, 'is interesting to remember, seeing that farmers are constantly complaining about the damage done by birds to their crops'. Evidently, at the time of his Lectures, farmers were advocating shooting birds to keep them under control. This, Chalmers Mitchell explains, is senseless, because all summer long, when most crops are growing, adult birds are busy collecting enormous quantities of insects and caterpillars of all kinds to feed their young. 'The result of killing off the birds would be at once a great increase in the number of insects.' The latter do infinitely more damage to crops and gardens than the birds (see Balfour-Browne's Lectures on insect life, page 37).

When it comes to working out how to feed young animals, Chalmers Mitchell has had plenty of first-hand experience throughout his time at London Zoo. He recounts his experiences of this to the RI audience, including feeding animals with some rather peculiar food. A young orangutan that refused all food, he says, was eventually tempted to drink milk flavoured with stewed rhubarb. On one occasion, he was brought a young tree hyrax (a small nocturnal mammal distantly related to elephants) that only ate sponge-fingers dipped in hot coffee or bread soaked in claret, but not port or champagne. In time the hyrax moved on to strips of toast soaked in milk and eventually took to its natural diet of green leaves.

He recommends giving young animals a chance to feed themselves as soon as possible and not to touch them while they eat; if you do, they're likely to snarl and bite because, in the wild, animals know they must defend their dinner from anyone else. In one instance, Chalmers Mitchell spent half an hour struggling with a young bear to try and feed it a teaspoon of castor oil. 'The keeper and I both got scratched and bitten and had our coats torn,' he says. They gave up and left a dish of castor oil in front of the bear, which then of course rushed up and greedily drank it all. 'Patience and experiment are the most successful methods with all animals,' he says.

'If you have a young alligator that's not eating,' he says, 'try immersing it in hot water.' When the cold-blooded alligator's body temperature is raised, it should soon become more lively and begin to eat. If that fails, he advises carefully poking bits of meat down its throat with a toothbrush. Members of the audience giggle, as Chalmers Mitchell reassures them this does the alligator no harm.

Sir Peter Chalmers Mitchell (1864–1945)

Born in Scotland, Chalmers Mitchell studied comparative anatomy at Oxford University. From 1903–35 he was secretary of the Zoology Society of London and under his guidance, London Zoo was transformed. In the *Leeds Mercury* on 30 December 1911, he said: 'Bit by bit we have replaced these old insanitary, dark and overheated buildings by lighter and airier structures.' Instead of keeping all the animals behind bars, some were kept in open concrete structures surrounded by moats. Chalmers Mitchell was a committed conservationist. Inspired by a visit to the Bronx Zoo in the US, he founded Whipsnade Zoo in Bedfordshire as a centre for conservation.

Even with parents (or adopted humans) providing food, animal childhood is not easy. An inescapable truth is that not all young animals will survive into adulthood. Chalmers Mitchell illustrates this with two very different species. He recounts how Charles Darwin worked out that if all young elephants survived, things would quickly get out of hand. That's despite the fact that they're among the slowest breeders in the world. A single pair of elephants should live for roughly a hundred years and produce six young. If all those offspring and subsequent generations survived, then over the course of 500 years that single pair and their descendants would give rise to some 15 million elephants. 'The world would very soon be so full of elephants it would be impossible to walk about,' says Chalmers Mitchell.

At another extreme is the turbot, a type of flatfish that produces 15 million eggs every year. Should they all reach maturity, the sea would be so teeming with turbot that 'people would not get seasick as they crossed the Channel,' he jokes. 'They could walk.' The audience laughs. The reason animals don't proliferate and take over the planet is because their young are good to eat (and often easier to catch), so get targeted by predators. A young chicken, Chalmers Mitchell says, is much tastier than a tough old rooster.

An idea Chalmers Mitchell has explored at the zoo is whether young animals are innately fearful of predators, snakes in particular. He re-runs some of his experiments in front of the RI audience, bringing in a large, live snake (a non-poisonous variety) and showing it to various animals to see how they react. Their responses are highly mixed. A yellow-crested cockatoo amuses the audience by apparently being terrified of a guinea pig, raising its crest and making a great fuss, but it seems unconcerned when the snake writhes and twists towards it. The Indian hill mynah, on the other hand, shrieks loudly and darts to the back of its cage. As soon as Chalmers Mitchell removes the snake, the bird immediately comes to the cage bars and pecks gently at his fingers. He's convinced the mynah knows about snakes and greatly fears them.

Most of the mammals he tests are indifferent to the snake. The guinea pigs and rats obliviously scuttle over it. A tree hyrax jumps backwards when the snake licks it with its long, forked tongue, but then reaches out and sniffs the reptile. It seems to realize the snake isn't good to eat and takes no further notice of it.

Next, Chalmers Mitchell shows the snake successively to a lemur, a very young capuchin monkey and a young baboon. The lemur was born at London Zoo, had never seen a snake until its visit to the RI Lecture Theatre and paid it no attention (there are no poisonous snakes in the lemur's home, the island of Madagascar, which could perhaps explain its indifference towards this one). The capuchin and the baboon, Chalmers Mitchell thinks, most

Sir Peter Chalmers Mitchell.

likely had no experience of snakes before they came to the zoo at a very young age. Their responses to the snake are dramatic. Both appear to be panic-stricken and the snake is removed at once. (Recent studies have uncovered evidence for a neural basis for some primates' apparently innate fear of snakes; particular nerves activate in a macaque's brain when it's shown a picture of a snake, even though it has never seen one before.)

As he turns to consider how the colours and markings of young animals often differ from adults', Chalmers Mitchell explains this could help to disguise them from predators at this early, vulnerable stage of life. To make his point, he brings into the

Lecture Theatre a range of stuffed birds, including a pair of regent bowerbirds from Australia. The adult male has splendid black and gold plumage while the juvenile, as reported in the *Manchester Guardian* on 3 January 1912, is 'a dull and shabby fellow'. (It makes sense that young and also female birds tend to have inconspicuous colouration, since it's only the mature males that need bright feathers to attract a female's attention during the mating season, and compete with other males.)

On the screen Chalmers Mitchell projects pictures of various other young animals with their parents, pointing out the differences between young and old. A young lion is covered in spots (like the live specimen that will make an appearance at the RI in Julian Huxley's Lecture in 1937; see page 53), which its parents have lost; the spots may help the young lion hide as it crouches in the dappled shade of grasses on the African savannah. And a young tapir is black with white stripes, unlike its dark, uniform parent; he suggests these patterns could break up the outline of the young tapir, making it difficult for predators to recognize it. All in all, early years can be a dangerous time for young animals and they need all the help they can get to make it through to adulthood. As Chalmers Mitchell says, 'The game of life is a game of hide and seek.'

A family of lions, with a male, female and spotted cub, from Chalmers Mitchell's book accompanying his Lectures.

Forecasting an invasion

In the summer of 1911, the year of Chalmers Mitchell's Christmas Lectures, New York was infested with millions of large, noisy insects with red eyes and orange wings. These cicadas had emerged from their seventeen-year youth as grubs underground. In his Lectures, Chalmers Mitchell predicted that New York would suffer its next invasion of cicadas seventeen years later in 1928, and he was indeed right, as reported that year in the *Yearbook of Agriculture*. The sudden insect influx stirred panic among members of the public even though, as the *Yearbook* reporter stresses, cicadas are not a serious pest. In fact, there was concern in 1928 that cicadas could become endangered because their habitats were being built on and paved over. 'Should the process finally result in the extermination of the cicada,' the reporter wrote, 'one of the entomological wonders of the world will have followed the dodo and the great auk.'

CHAPTER 2

THE HAUNTS OF LIFE

Sir John Arthur Thomson

1920

A splendid journey through the living world awaits us in these Lectures, as Thomson leads us through six great 'haunts' of life. The term *haunt* has fallen out of use and today we might refer to biomes, great swathes of the planet where particular organisms live. Along the way Thomson reveals how species have adapted to meet the particular challenges of life in water, on land and in the air. He gave these Lectures almost a hundred years ago and while advances have clearly been made (in particular in deep-sea exploration and our understanding of the origins of life), they reveal that much was known even then about the details and wonders of life on earth.

'Life is like a river that is always overflowing its banks,' says Thomson as he introduces his Lectures, describing how living things have spread across the planet, from deep oceans to mountaintops, from pole to pole, and everywhere in between. 'One may almost say that over the earth and sea life is omnipresent,' he says, 'but it is very useful to distinguish six great haunts of life.'

The first of Thomson's haunts will be familiar to many – the seashore – but nevertheless it still holds many surprises. It is where polar bears hunt and walrus bask, where sea snakes and sea turtles come to lay their eggs. It's the home of the curious-looking anglerfish that shuffles along the shallow seabed on its fins and dangles a lure, a fleshy growth on its forehead that looks like a wriggling worm, tempting other fish into its enormous mouth (deep-sea anglerfish have similar lures that glow in the dark). It is, Thomson points out, a fish that 'really fishes with a fishing rod'.

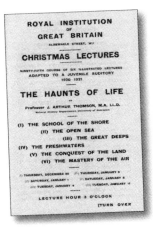

Lecture programme (front cover).

A drawing specially commissioned for the Lectures by London Illustrated News, *depicting Thomson being besieged by his youthful audience as he shows a stuffed manatee.*

But the shore is a difficult place for life to thrive. 'It must be an interesting and stirring place to live in,' he says, 'but no one could call it easy.' There are waves and storms to contend with and on a daily basis creatures must cope with being flooded, then exposed, in dry air. It's also a crowded place, crammed with species. 'Almost every kind of creature is there,' says Thomson.

As our Lecturer dips his toes into this first haunt, his great flair for shifting his focus and telling stories is in full effect. One moment he's concentrating on tiny, fascinating details of individual animals – the graceful kicks of a barnacle's feathery legs as it sifts the water for food, or the crabs that camouflage their shells with scraps of sponge and seaweed – and the next he steps back to paint a broad picture of how these creatures interact and compete with each other.

Then he dives into the open sea. 'Not a bad subject for a New Year's Day,' he begins in his second Lecture on 1 January 1921, 'for we might well wish one another a more open sea, away from the rocks, and into a place of freedom and spaciousness.' Compared to the crowded, ever-changing shore, living in the open sea is much less of a struggle.

To introduce his audience to this haunt, Thomson projects pictures on the screen of what he calls 'floating sea-meadows', vast tracts of water filled with minute algae (generally known today as phytoplankton). 'On these everything else depends,' he says. It leads him to point out an unexpected link. 'What connection,' he asks, 'is there between the amount of sunshine in the spring months and the supply of mackerel at Billingsgate market?' The mackerel, he explains, feed on copepods (microscopic crustaceans), which in turn feed on phytoplankton that use chlorophyll to harness the sun's energy. It follows that the more sunshine there is, the more 'sea soup' there is for the mackerel. 'And so the world goes round,' Thomson says. 'Nothing is ever lost; all things flow.'

Animals of the open sea spend their whole lives either actively swimming or drifting. As Thomson says, this isn't a place of rest. Among the swimmers there are sea snakes, herring, squid and also marine mammals, which he describes as having a 'great bundle of fitnesses', his term for the many ways they're well adapted for life in the ocean. He clearly admires the whales. 'All of them have such mastery of their medium that they must be ranked among the conquerors of the open sea,' he says. Their bodies are highly streamlined, they keep warm with thick layers of blubber and sift small animals from the water using dense bristles in their mouths, known as whalebone or baleen (which was used to make women's corsets). 'Unfortunately for the whale,' Thomson says, 'man long ago discovered the value to himself of the whale bone and the blubber.' At the time of Thomson's Lectures, commercial whaling was still in full swing and, as he points out, due to the modernization of fishing vessels and harpoons 'this interesting animal is fast disappearing from the seas'.

Turning to the drifting sea creatures, Thomson charms the audience with his description of an unusual type of octopus called the paper nautilus or argonaut. The most remarkable thing about them, he says, is that the female makes 'the most beautiful cradle in the world' to carry and protect her eggs. He describes how the little octopus secretes a calcium carbonate shell from webs on the ends of two arms (he says argonauts drift on the surface in their shells, but we now know they swim actively underwater by jet propulsion). Another ocean-going mollusc is a snail with a violet shell, which secretes a float of frothy bubbles.

From the open sea Thomson now sinks down to explore the wonders of 'the great deeps'. Until a few decades before his Lectures, it was widely thought that nothing lived in the deep sea. Nevertheless, when scientists began studying this inaccessible haunt in detail, such as in the groundbreaking expedition of HMS *Challenger* in the 1870s, they used trawl nets and

Sir John Arthur Thomson (1861–1933)

Born in Saltoun, East Lothian, Sir John Arthur Thomson was a naturalist and expert on soft corals. From 1893 to 1899 he taught at the University of Edinburgh before taking up the chair in natural history at the University of Glasgow where he stayed until his retirement in 1930. He was a prominent popular science writer and penned many books including *The Evolution of Sex* (1889), *Science and Religion* (1925) and *Outline of Science: A Plain Story Simply Told* (1922), which sold more than 100,000 copies.

dredges to bring up extraordinary wildlife from the depths. With an average depth of 2.5 miles (3.7 km), and 'deeps' of over 6 miles (9.7 km), exploring the abyss is extremely challenging. 'No one has ever seen it,' Thomson says. (It would be another decade before people ventured into the depths, as we'll see in Julian Huxley's Christmas Lectures, page 53.) 'The biggest fact is that there is no "deep" too deep for life,' he says.

This haunt poses a unique set of challenges for life. In Thomson's view, this is 'a deep, dark, cold, calm, silent, monotonous world'. The lack of sunlight means there's no plant life and food is hard to come by. Many animals are adapted to eat each other and to swallow whatever comes along, no matter how big. Some deep-sea fish are 'nothing but a mouth and a stomach', he explains. The only other food source are the dead animals and plants, including plankton, that sink down from above 'like snowflakes on a quiet winter evening'. (Later in the twentieth century, another vital food source in the deep sea was discovered in the form of bacteria that harness energy not from sunlight but from chemicals, so-called chemosynthesis.)

Sea lilies (crinoids) growing from the deep-sea floor.

Thomson describes for his audience the deep-sea animals that are found nowhere else on earth, like the sea-lilies, relatives of starfish that live in great beds 'like daffodils by the lakeside'. And while there's no sunlight reaching the deep, it isn't in fact completely dark down there. Via chemical reactions, many species of fish, corals, crustaceans and jellyfish are bioluminescent, using their own light to communicate, search for prey and warn off enemies in the deep. Thomson imagines what this would look like: 'In the absolute darkness of the abyss they would appear as ghostly silver-blue shapes, glimmering like an electric lamp through dense fog on a dark, moonless night.' There are also things missing from the deep sea that are common elsewhere. Thomson tells the audience that this haunt contains no bacteria. 'That means there is no rotting.' (The deep sea is in fact crawling with bacteria, but special techniques are needed to keep them alive at the surface; now scientists are examining these bacteria for unusual molecules, including potential new antibiotics.)

'Where did the deep-sea animals come from?' Thomson asks. In his view, life most likely began in shallow seas at the shore and then spread from there. Shore-animals could have migrated gradually into the abyss, following the drifting food, he suggests. (While life did indeed first evolve in the oceans, it remains unclear exactly where; a recent theory suggests life began on deep-sea hydrothermal vents, ecosystems that were only discovered in 1977.)

Leaving the deep sea, Thomson next focuses his attention on freshwaters. Ranging from deep lakes and small ponds, to 'great rivers and purling brooks, swift torrents and sluggish streams', these cover less than a hundredth of the earth's surface but nevertheless contain a great variety of life.

Freshwaters are dynamic and shifting, and many inhabitants of this realm come and go, spending only part of their lives in fresh water. Eels spawn in the Sargasso Sea, in the mid-Atlantic, and migrate to rivers and lakes where they feed and grow. Salmon undergo immense migrations between salt and fresh water. Thomson brings into the Theatre the stuffed body of a young manatee, or sea cow, a few feet long. This, he tells his young audience, is typically a coastal animal but it goes far up rivers and is now found, for instance, in the everglades of Florida. But this fluctuating realm also brings dangers. There is always the risk, for example, that ponds and lakes will dry up, especially in warm countries. In response, animals have evolved strategies to sit out parched times. Thomson describes the strange 'mud-fish' (known now as lungfish), from tropical Africa. When waters get very low, they dig a burrow in the mud that dries and hardens and the fish lie safely inside until the rainy season arrives. Fish encased in mud have been brought back to England and wake up even after many months. 'A fish out of water indeed!' Thomson jokes. Another great risk – in streams especially – is flooding. Some freshwater animals, like leeches and insect larvae, have suckers and gripping organs to prevent them getting swept away after a rainstorm.

Thomson considers how animals came to live in freshwaters. He describes how animals from the shore might have slowly moved up rivers away from the tide's influence. Some animals have come to freshwaters from land, like the water-spider. Thomson fondly describes the female spider spinning a web at the bottom of a pond, anchoring it with threads and filling it with air from the surface, trapped in hairs on her body. Then, once her web is full of air she lays her eggs inside. 'This spider,' Thomson explains, 'found an empty corner – an empty niche of opportunity – full of difficulties, to be sure, but offering new opportunities of food and safety.'

Next we dry ourselves off to explore the fifth of Thomson's haunts and follow the stories of aquatic animals that embarked on three great invasions of land and led to life on earth as we know it today.

First came the worms. 'For a long time they had the whole realm to themselves with no enemies,' he says. The worms helped make conditions hospitable for other organisms by eating and breaking down the simple plants that were also emerging onto land and in the process forming fertile soils. Following the worms were arthropods, the 'joint-footed' animals. Thomson introduces *Peripatus*, 'a beautifully coloured soft-bodied animal, worm-like in shape, but with simple stumpy limbs', and suggests this is what ancient arthropods looked like (also known as velvet worms, they live across the world, especially in rainforests, and are now considered to be close relatives of arthropods). Arthropods in this second great invasion included centipedes, millipedes, spiders and insects, which soon formed partnerships with plants, carrying pollen between flowers (we'll hear more about the relationships between animals and plants in Sue Hartley's Lectures; see page 177).

A velvet worm (Peripatus), *from Thomson's book.*

Then, at the end of the Devonian period (around 359 million years ago), the amphibians made their appearance, which eventually led to reptiles, birds and mammals (a series of remarkable fossil finds has since revealed how particular fish evolved to live on land giving rise to all the tetrapods, the vertebrates with four limbs, including humans). This great invasion, Thomson reveals, is still going on. He shows specimens of a fish named *Periopthalmus*, otherwise known as mudskippers, which spend much of their time out of water, clambering on roots and trunks in mangrove forests. Plants have given animals many great benefits, he says, including providing trees to climb on. An arboreal lifestyle 'opens up new possibilities of movement, of feeding, or nesting'. And all sorts of animals do it, from green and agile tree snakes, to crabs, insects, monkeys and birds. At the end of his stories of land invasions, Thomson leads us to a perfect jumping-off point for his final Lecture, as we follow the animals that leap from treetops.

'The last haunt to be conquered was the air,' says Thomson. Flying is a means of escape from predators, he explains; just think of a cat watching a sparrow fly out of reach. Also, he says, 'it has led to an annihilation of distance and to a circumventing of the seasons'. Birds undertake long journeys between breeding grounds and feeding grounds, often in distant parts of the planet. Golden Plovers of Hawaii seem to think nothing of setting out for Alaska.

Thomson introduces the four groups of animals that fly: insects, birds, bats and the 'flying dragons' – the long extinct pterodactyls (strictly known today as pterosaurs). Each flies their own way, flapping, hovering and sailing through the air. Then there are the animals that make attempts at flight and are, he says, 'splendid failures'. Flying fishes don't actually fly, but leap into the air and glide along without flapping their 'wings', fashioned from wide fins. There are tree-frogs with webs between their toes and flying phalangers (also known as flying lemurs) with a flap of skin stretched between their front and hind legs; they leap from trees and glide along, assisted by their inbuilt parachutes (we'll hear more about how wings evolved in Richard Dawkins's Lectures; see page 120).

As he draws his Lectures to a close, Thomson calls

attention to animals that symbolize 'a quality of endeavour and experiment, or insurgence and adventure' that we see through all the haunts of life. These are the gossamer spiders, terrestrial creatures that fly without wings. In autumn, they climb gateposts and tall plants and let out long threads

Gossamer spiders on their aerial journey, from Thomson's book.

of silk, which catch the wind and whisk them away, perhaps to somewhere with more space and food. The spiders have attempted the apparently impossible and achieved it, filling us, Thomson concludes, 'with a reasonable wonder at the adventurousness of life'.

From the archives . . .

Specially commissioned drawings were published in the *London Illustrated News* in January 1921 to accompany Thomson's Lectures, beautifully illustrating the six great haunts of life.

The 'school of the shore'.

CHAPTER 3

CONCERNING THE HABITS OF INSECTS

Francis Balfour-Browne

1924

Balfour-Browne takes us into an intricate, hidden realm as he explores the complex lives of insects. He shares his fascination with butterflies, bees, beetles and dragonflies, and shows how insects pass through radically different stages of life, undergoing complete metamorphosis, radically reorganizing their bodies and often shifting between different 'haunts', first swimming then flying. We learn how insects affect our everyday lives, through impacts on health and food production, and gain a sense of the surprising things we can see if we learn how to look into the insects' world.

✧

'With many people, collecting insects is merely an incidence, like the measles.' So begins Balfour-Browne in the first of his Christmas Lectures about the wonders of the insect world. 'They catch the complaint from someone else and usually recover,' he says. But some people never get over this affliction, including Balfour-Browne himself. Over the course of his Lectures he hopes some of his audience might succumb to the charms of insects and insect collecting.

During his first Lecture, on the table in front of him sits an aquarium tank with water beetles 'groping about', as a reporter from the *Manchester Guardian* reported the following day. These large,

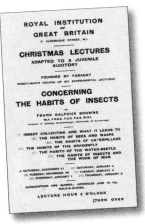

Lecture programme (front cover).

dark and shiny beetles are the insects Balfour-Browne has come to know the best. His searches for them across the length and breadth of Great Britain – some 350 species in total – have led him to learn not only about the lives of beetles but also about the living world more broadly.

He introduces his audience to the idea of ecological communities, in which different species coexist by having different habits. He describes a pond community where squeak beetles live in the fine oozy mud at the bottom and great diving beetles reside among weeds, rarely entering the mud – and so the two species can share a pond without getting in each other's way.

Long before anthropogenic climate change was recognized, and pre-empting the popularity of members of the public helping to collect data as so-called 'citizen scientists', Balfour-Browne advocates the use of amateur insect collections for recording changes in ecosystems. By mapping out the places where certain water beetles live at certain times, and the habitats they occupy – from tide pools to peat bogs – he shows that species from continental Europe have swept northwards, as temperatures increased since the last ice age. At the same time, other species that prefer cooler conditions have shifted west and north. Insect collections not only tell us about the past, he says, but 'they are going to be of use in the future', as records of where insect species have been found will help scientists track how ecosystems shift and respond as our climate continues to warm.

In each of his next four Lectures, Balfour-Browne focuses on a particular group of insects in turn. First are bees and wasps (which belong to the group of insects called the Hymenoptera). Many species are social and live together in hives, but Balfour-Browne concentrates on solitary species. He tells the audience about the 'bee wall' he built to study the solitary bees and wasps that live in his small garden in Cambridge. He made it from ventilation bricks with a short glass tube pushed into each hole (a similar idea lies behind 'bee boxes' sold today in garden centres although these are without the glass tubes, which Balfour-Browne added so he could carefully pull them out and inspect what was going on inside).

A common visitor to Balfour-Browne's wall was the Red Osmia. 'This bee reminds one of a small humble-bee,' he says (humble-bees are better known today as bumblebees), 'with a dense coating of reddish brown hairs.' In the spring, female Red Osmias would claim glass tubes and begin busily building walls inside them, from mouthfuls of soil. They then fly off to find flowers and bring back pollen and honey, which they added to the tube before laying a single egg and sealing it in with more soil. The females repeat this process, filling the tubes

A 'bee wall' showing house bricks with glass tubes (top) as used by Balfour-Browne to study solitary bees and wasps. Some of the bees that visited his wall (bottom) included the Red Osmia (1), the Leafcutter bee (2), the Anthophora bee (3) and cuckoo bees (4, 5).

with more soil, honey and eggs. Ten days later the bee larvae hatch and tuck into the food left by their mother. The growing larvae are at first small wriggling grubs, before they wrap themselves in a silk cocoon and form a pupa, a white creature with legs that begins to look like an adult bee. Eventually, in the autumn they transform one last time, bite through the mud walls of their birthing chambers and fly off as mature bees. All of this Balfour-Browne learnt while carefully watching his bee wall for some months.

Various solitary wasps also visited the wall. The main difference between bees and wasps is the larvae's food. Instead of pollen and honey, female wasps stock their nests with spiders, caterpillars and flies. The mothers sting and paralyse the prey to keep them alive and fresh, while making sure they can't walk off or fly away. It's another example, Balfour-Browne says, of similar species coexisting in a close community. In this case bees and wasps get along without competing for the same food: bees are vegetarians, wasps are carnivores.

Caterpillars, the subject of the third insect Lecture, have been described by some as 'upholstered worms'. Of course they're not worms at all, Balfour-Browne points out, but the larval stage of butterflies and

---

Francis Balfour-Browne (1874–1967)

Generally known as Frank, or by the nickname B-B, Francis Balfour-Browne was born in London and studied botany at Oxford University. After a brief stint working in law, he returned to Oxford to study zoology and became an expert in water beetles. In 1902 he became director of the Sutton Broad Laboratory in Norfolk, Britain's first freshwater research station. Balfour-Browne taught for a few years at Queen's College, Belfast, before moving in 1913 to Cambridge University. In 1925 he took a part-time chair in entomology at Imperial College London, before retiring in 1930 to devote himself to the study of his beloved water beetles. He's remembered in the Balfour-Browne Club, an international water beetle conservation trust.

---

moths. Rather than drawing attention to the colourful adults, he instead zeroes in on these younger forms and in particular their spit. 'Silk, as produced by the caterpillar, is the product of the salivary glands,' he says. It's a liquid that becomes solid on contact with air and caterpillars make a lot of it. 'All silk-producing caterpillars are perpetually dribbling,' he says. (Sue Hartley explores the chemical wonders of caterpillar spit in her Lectures; see page 177.) As they move forwards, the silk flows from spinnerets on their lower lip, forming a thin thread. And caterpillars use silk in many different ways. 'While the caterpillar feeds,' Balfour-Browne says, 'the oozing saliva is swallowed with the food and assists in its digestion.' Caterpillars commonly cover plants in nets of silk, giving them a foothold while they're eating leaves. He describes how caterpillars of Little

Balfour-Browne showing his young audience his insect collections after one of his Lectures.

Ermine moths string webs across hedgerows until it looks as if fine muslin has been hung out to dry.

A danger all caterpillars face is falling off a plant, which Balfour-Browne says, 'may entail a long and weary walk back to the feeding place'. The silk trail forms a lifeline, like the climbing ropes mountaineers use, which the caterpillar can easily climb back up. They even have a braking system in their spinnerets to halt their fall. Trails of silk can also help caterpillars find their way back to their nest after they've wandered off to find food.

From these lines and webs, Balfour-Browne traces the development of silken shelters, which caterpillars build by gluing together bits of stick, leaf stalks, lichens and grass. He tells of Hammock moths from Brazil with strange shelters that they carry around, enlarging them as they grow. 'This structure is composed of the excrement pellets of the caterpillar bound together with silk,' he says.

As an article in the *Observer* on 4 January 1925 reported, Balfour-Browne took on 'the phenomenally patient role of foster parent of a family of dragonfly nymphs'. The Lecturer tells his young audience how, seventeen years previously, when he was living in the Norfolk Broads, he raised 175 dragonfly nymphs (the term for their aquatic larvae), which he hatched from eggs collected in the fens, housing each one in a separate glass tumbler of water. At the time of these Lectures very little was known of the dragonfly's life cycle and Balfour-Browne painstakingly filled in many important details. 'Every day for a fortnight I measured each one of these nymphs,' he says. 'I started early in the morning and continued till late at night.'

Among his many discoveries, Balfour-Browne found a possible explanation for a striking phenomenon he calls 'dragonfly storms', when millions of adult dragonflies fill the air and festoon trees in glistening curtains. Sporadic swarms like this, he says, have hit South America and Heligoland – a small archipelago in the North Sea – perhaps following mild winters. In his experiments at home Balfour-Browne kept some nymphs over winter in a warm incubator

Francis (Frank) Balfour-Browne.

and found they matured into adults after only ten months, at least a year sooner than the nymphs kept outside. Nymphs in the wild could also speed up their development during a warm winter and emerge at the same time as generations two or three years older, which join together to form large swarms.

According to an article in the *Observer* the following day, the audience enjoyed Balfour-Browne's description of the way dragonfly nymphs 'split their skins and sailed out on their virginal flights as free, glorious spirits of the air'.

Next Balfour-Browne turns the spotlight on water beetles, the group of insects he's devoted much of his life to studying. 'The point that struck one about the Lecture was that there was nothing in it taken second-hand from books,' said a reporter from the *Manchester Guardian* on 7 January 1925. 'If Mr Browne is tracing the life history of the Lapland Dysticus, or the silver beetle, he describes exactly what he has discovered for himself in a very delicate and difficult series of observations.'

Balfour-Browne explains how he studied water beetles in small ponds made from wooden barrels sawn in half. Into these he introduced a rare beetle species that he collected on the islands of Skye and Eigg, off the west coast of Scotland. A month later there was no sign of life and he assumed they'd all died or escaped. But when he

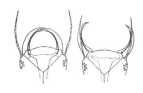

Diagram of the head of a water beetle larva.

emptied out the ponds he found the beetles buried in the mud at the bottom and realized they were in fact asleep for the winter. When his beetles woke up in March, he watched the females drill holes in water plants to lay their eggs, which in time hatched into voracious larvae. The larvae grab prey in their formidable jaws, inject them with digestive enzymes and suck the juices out through tubes in their

Drawings of the feeding habits of water beetle larvae. Eating a pond snail (left) and eating another larva (right).

mouths, leaving just their prey's empty, crumpled skin. He gives a graphic description of the greater silver beetle, a species with specialized jaws that act as a can opener to break into the shells of pond snails. And great diving beetle larvae are cannibals, he says, that 'have no respect for one another and four placed in a large tub were quickly reduced to one'.

Balfour-Browne captivates the audience with a film of water beetle larvae feasting on tadpoles. As the *Manchester Guardian* reporter wrote, this 'gave a vivid impression of the liveliness and ferocity of the larvae when it hunts and plays havoc among an innocent school of tadpoles. This picture was a revelation of the remorseless struggle for existence that goes on under the placid surface of every pond.'

Balfour-Browne brings his Lecture series to a close by considering the influence insects have on the human world. They provide useful products like silk, honey (especially back in the days when people drank a lot of mead), cochineal dye and shellac for making lacquer and vinyl records (today, there is even talk about insects being the protein-rich food of the future). Insects also impact on human health and agriculture, and Balfour-Browne gives several cases where knowledge of insect life cycles and communities have helped combat problems.

For example, by the turn of the twentieth century, links had been made between blood-sucking mosquitoes and the transmission of both malaria and yellow fever. Steps were then taken to combat the diseases by targeting specific life stages of the mosquitoes, a group of flies with aquatic larvae. In the Cuban capital, Havana, a notorious yellow fever hotspot, a campaign in 1900 aimed to eradicate water bodies where mosquitoes breed, including blocked rain gutters and open containers. Puddles were either filled in or covered in oil, forming a layer that

Balfour-Browne shows live water beetles to his audience at the end of one of his Lectures.

mosquito larvae can't pierce with their breathing tubes, effectively drowning them. Such measures quickly led to a drop in yellow fever outbreaks.

Knowledge of insects can also help protect crops from pests. Early or late sowing can avoid times when insects are most abundant, in particular the hungry, leaf-eating caterpillars. In Egypt, Balfour-Browne says, cotton plants are forced to mature early by withdrawing irrigation, so the crop can be harvested before the arrival of moth larvae, known as pink bollworms, which wreak serious damage. Insect studies may also reveal which plants the pest species prefer to eat. A crop can be saved from attack by planting it among occasional rows of the favoured plant, known as a 'trap crop'.

Insects only break out into pest proportions when the 'balance of nature' is disrupted, says Balfour-Browne. And the problem is that 'now man is always interfering'. People destroy wild habitats and replace them with large areas of crops, which supply insects with easy food and their numbers explode. 'If nature were given time, she would undoubtedly restore the balance,' he says. 'But as long as man continues to cultivate land, nature is not given a fair chance.' Balfour-Browne also tells his audience about an experiment in French orchards to get rid of apple-blossom weevils that destroy flowers and ruin crops. Infected blossoms were picked from the trees and kept in mesh boxes that trap the weevils while allowing smaller insects to fly in, which parasitize the pests. After ten years of this treatment the weevils were being kept in check by the boosted population of naturally occurring parasites (we'll hear more about insect parasites in Sue Hartley's Lectures; see page 177).

Rather grudgingly, it seems, Balfour-Browne admits that increasingly experts are being employed as 'economic entomologists' to help control pests. 'I deplore the suggestion that anyone might take up entomology with a view to making a living,' he says. In his view, such research should not be driven by commercial interests. 'The study of insects justifies itself,' he says, 'and however useless it may appear to be, its interest alone will repay the student and that is the spirit of the man of science' (and presumably the woman of science also).

CHAPTER 4

RARE ANIMALS AND THE DISAPPEARANCE OF WILD LIFE

Sir Julian Huxley

1937

The human impact on the living world is well known today and we often hear about climate change, overfishing and other environmental problems. But you might be surprised to learn from Huxley's Lectures that concerns about wildlife and the environment are not new. Yet Huxley also kindles a sense of optimism, highlighting that there's much we can do to protect wildlife and prevent more extinctions in the future.

The star of Huxley's first Lecture is without doubt an eight-month-old lion cub called Max. Newspapers across the country reported on this unusual visitor and how uncomfortable he clearly felt being in the limelight. 'The children sat in breathless silence', according to an article in the *Daily Telegraph* on 29 December 1937. Huxley had already warned the audience not to clap. 'He is a little temperamental,' he warns.

The reason for the lion's visit to the Lecture Theatre is to show the spots on his belly. Huxley tells the story of the spotted lions, or marozi, that some zoologists claimed were a rare variety from the mountains of Kenya. Their spots could have camouflaged the lions in the montane forests. Young lions, like Max, usually have spots, then lose them as they grow up. Maybe there are lions that keep their spots, Huxley suggests, but none have been seen alive (still today there have been no confirmed sightings). Like a naughty dog, the little lion squats on his haunches and slithers along the polished floor, pulled along on his lead by his keeper. At the first chance he gets, Max makes a bolt for the door.

Plenty more excitement lies in store as Huxley introduces various rare and splendid animals, all on

Huxley introduces Max the spotted lion cub to his audience.

loan from London Zoo. Flapping its wings in a cage is a large bird of prey, which the children in the audience have probably never seen in the wild; it's a red kite from Egypt. At the time of Huxley's Lectures there were thought to be only four breeding pairs left in Britain. But this hasn't always been the case. Huxley tells of the red kites in the Middle Ages that scavenged offal from outside butchers' shops in London. In more recent times, they were almost wiped out because people collected their eggs. (Following a long-running reintroduction programme, today there are many hundreds of red kite breeding pairs in Britain.)

The rarity of some animals, Huxley says, stems from their isolation in remote parts of the world, including many islands. To demonstrate, he brings out some strange mice from the Faroe Islands with longer, bigger ears and bigger hind feet than ordinary house mice (recent genetic studies revealed these mice were introduced by people centuries ago, from Britain and Scandinavia, and have since evolved into distinct subspecies). He also brings in a pair of parrots with striking green and blue feathers. One is an endemic species that only lives on the Caribbean island of St Vincent, and has a bright yellow crown on its head. The other, from the mainland, lacks the distinctive crown, demonstrating how small differences can arise when species are split and isolated on islands (today, the Saint Vincent parrot is considered vulnerable to extinction due to deforestation and hunting).

As well as live zoo animals, Huxley uses a magic lantern (an overhead projector of the time) to project images onto a screen of a menagerie of other rarely seen creatures and takes his audience on a pictorial exploration of the Galapagos Islands.

Huxley (left) shows his audience a rare Caribbean parrot.

We meet marine iguanas, flightless cormorants, dozens of ground finches and the giant tortoise; all these unusual animals, which live nowhere else, helped convince Charles Darwin that new species can evolve when populations are split, in this case when a few intrepid animals made the long journey, flying or floating from the mainland and ending up isolated in the Galapagos.

Huxley also uses the magic lantern to illustrate stories of rare creatures that never actually existed, but have some basis in reality. He shows depictions of sea serpents but tells the audience there is every reason to believe reported sightings, including from two fellows of the Zoology Society of London. What these people saw, Huxley thinks, weren't in fact snakes but giant 'cuttlefish' (species of mollusc known nowadays as giant and colossal squid). Other possibilities are basking sharks, pods of dolphins and ribbonfish (also known as oarfish), which grow up to 11 m (36 ft) long and could easily be mistaken for a mythical serpent.

He also relates the legend of a tree with leaves that drop off and run around on little legs and, when trodden on, ooze blood. 'Everyone laughed at this story', reported the *Scotsman* newspaper on 31 December 1937. 'But there is no doubt,' says Huxley, 'that it is based on one of the most extraordinary creatures in the world.' He places some objects on the magic lantern that look like leaves. 'Under the light they uncurled and

*Lecture programme
(front cover).*

proved very much alive indeed', wrote the *Scotsman* reporter. These leaf insects camouflage themselves so that many animals won't bother eating them.

Stories of dragons, Huxley thinks, originate in ancestral memories of extinct ice-age beasts, like cave lions and hyenas. The closest to a real dragon is the world's largest lizard, the Komodo dragon. Huxley hasn't brought one with him from the zoo. 'They are extremely fierce and difficult to box up,' he admits. He does, however, open up a case that he says contains dragon bones. In fact they're ancient bones of three-toed horses, which are sold in China as 'dragon bones' to make into traditional medicines. These horses roamed in great herds across the plains of Asia 10 million years ago but went extinct, probably through changes in climate.

Huxley also tells his young audience about the strange inhabitants of the deep sea, as he reveals the obscure animals that are rarely found because they live in inaccessible places. He shows pictures of deep-sea fish with rows of red and blue 'headlights' along their bodies, which they use to lure in smaller fish before swallowing them. (It's now known that deep-sea animals use bioluminescence for various purposes, including camouflage, defence and attracting mates.)

In a story from forty years previously, Huxley tells of the Prince of Monaco, a keen zoologist, who was sailing his yacht when he encountered whalers killing a sperm whale. The whale vomited and the prince dispatched a boat at once to collect some of the sick. It contained remains of 'cuttlefish' (again, probably giant squid), which the whale had eaten down in the depths and were still sufficiently intact to identify as two species that were new to science.

At the time of Huxley's Lectures, deep-sea exploration was still in its infancy. In the early 1930s, William Beebe and Otis Barton were the first people to observe deep-sea creatures in their natural habitat, through the windows of a cramped metal sphere called the Bathysphere. They set a world record in 1934 for the deepest-ever dive to 923 m (3,028 ft) which Otis eventually beat in 1949. (Modern deep-sea submersibles regularly dive down thousands of metres; in 1960, Don Walsh and Jacques Piccard were the first people to reach the deepest point in the oceans, 10,911 m (35,797 ft) down at the bottom of the Challenger Deep in the Mariana Trench.)

Moving on from deep-sea exploration, Huxley next takes us on a journey into the past, to a time when giant reptiles wandered the earth. He shows pictures and models of various dinosaurs and, to demonstrate just how long ago they were alive, he holds up a ream of typewriter paper. The pile is an inch thick (2.5 cm) and contains 480 sheets. Huxley says that if each sheet represents a thousand years, then the ream corresponds to the time that's passed since the beginning of the last ice age. To mark the time since the dinosaurs went extinct (roughly 65.5 million years ago) the pile of paper would be 3 m (10 ft) high. Go back even further, to the time when trilobites scuttled over the seabed (up to 520 million years ago) and, as Huxley tells us, we would need a pile of paper as tall as the Eiffel Tower. In all that time, life on earth has been through what Huxley calls 'successive exterminations' when multitudes of species went extinct. He introduces living examples of rare animals that have mysteriously survived through aeons of mass extinctions. From an aquarium tank he fishes out a king crab (known, these days, as horseshoe crabs), a rare survivor from a group that disappeared long ago. 'This used to be thought of as a curious type of crustacean,' Huxley says, 'but it was much more nearly related to extinct sea scorpions.'

Another example of a rare survivor is the flying phalanger from Australia, which is brought in to demonstrate its unusual abilities. 'The keeper flung it high in the air,' wrote a reporter from the *Scotsman* on 5 January 1938; 'it spread out its membranes – rather as if it were holding wide its coat to catch the wind – and glided gracefully down into his hand again. An encore to this feat was insisted upon.'

The audience give an enthusiastic ovation when Huxley projects a 12,000-year-old cave drawing of a wild horse, described by the *Scotsman* newspaper as a 'fine, virile, lifelike looking animal with a twirling tail'. Huxley reveals that these same wild horses still exist although only in captivity. (Today a few Przewalski's horses – the world's only undomesticated horses – have been released back into the wild, including in the evacuated area around Chernobyl.)

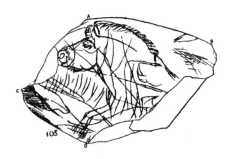

Cave drawing of a wild horse that used to be common in Europe, from Huxley's Lecture programme.

In his penultimate Lecture, Huxley turns his attention to the impacts people inflict on the living world. Although these Lectures were given eighty years ago, he takes a thoroughly modern view on endangered wildlife and the urgent need for conservation. 'Once a species is gone, it is gone for ever', he reminds the audience. 'The most extraordinary example of extinction,' Huxley says, 'was that of the passenger pigeon of North America.' He describes the enormous flocks, a mile wide containing two billion birds, which would take hours to pass overhead. They were slaughtered for food and just twenty-three years before this Lecture, the last passenger pigeon died in Cincinnati Zoo (her name was Martha).

Among the animals that survive but are threatened with extinction, Huxley mentions the brilliant blue butterflies that are slaughtered in huge numbers to meet a craze for butterfly jewellery (whole butterflies or parts of their wings were set in pendants and brooches). Twisting and wriggling in an aquarium that has been brought into the Theatre is a pair of foot-long alligators from China. This species, Huxley says, is nearly extinct in the wild because of deforestation, pollution and habitat destruction of their native rivers in China (the Chinese alligator remains critically endangered today). The zookeeper also brings on an eight-foot python whose skin is in high demand for making leather ties and shoes. At the end of the Lecture, the children are invited to meet the python which, according to an article in the *Manchester Guardian* the next day, they did with 'utmost intrepidity and only the occasional grimace, as the long muscular body . . . coiled around their necks and arms. The python, for its part, bore the handling with extreme good humour.'

Certainly a much more troubling scene awaited the audience in Huxley's final Lecture, which begins with a film of seal pups being clubbed to death. 'Unless seals are protected there is no chance for them to survive,' he laments, describing the trade in their newborn white fur (seal hunting continues today, although under tighter regulations). The same business, but on larger scale Huxley points out, goes on with the harpooning of whales (commercial whaling continued for many more years, but eventually came to a halt as we'll see in David Attenborough's Christmas Lectures; see page 100).

Sir Julian Huxley (1887–1975)

Born into the distinguished Huxley family, Julian Huxley's grandfather was biologist T. H. Huxley, a great supporter of Charles Darwin, and his brother was *Brave New World* author Aldous Huxley. He studied zoology at Oxford University and afterwards conducted pioneering studies of evolution and embryology in North America, at Oxford University and King's College London. He was the RI's Fullerian Professor of Physiology from 1927–31, secretary of the Zoological Society of London from 1935–42 and in 1951 co-founded the World Wildlife Fund. He was a great communicator of science and exponent of wildlife conservation and wrote many books, including *The Life of Science,* co-authored with H. G. Wells.

'What can we do,' Huxley asks, 'to check this tendency to destruction of animal and bird life through sheer greed and ignorance on the part of human beings?'

One solution he strongly supports is setting aside special areas where wildlife is strictly protected – in small sanctuaries, larger reserves and big national parks. He plays films showing abundant wildlife in national parks in Africa and Canada: lions, cheetahs, baboons, giraffes, grizzly bears and deer. The Lecture Theatre also echoes to the sound of wild kittiwakes. These gulls were recorded in one of the reserves around the British Isles that protect birdlife. 'We ought to have many more national parks,' says Huxley. (Since his Lectures, the protection of the living planet has increased. In 2014 reserves covered 15.4 per cent of land and freshwaters and 3.4 per cent of the oceans. Many experts think at least a third of both realms needs to be protected.)

Huxley recounts other conservation success stories. A spotlight in the Theatre beams on a goldfinch in a cage, with its striking white, black and red head and yellow striped wings. 'When I was a boy,' Huxley says, 'it was a rare thing to see a goldfinch, but happily, we could now hardly go into the country without seeing one.' Goldfinches were brought back from the brink by the Wild Birds Protection Act, which tightened laws on trade in feathers and egg collecting. He advocates campaigns that help shift public opinion on environmental issues. In particular, he suggests 'pressure can be brought on commercial and sporting interests to persuade or force them not to kill the goose that lays their golden eggs. This is very necessary with whales, fur-bearing animals and migratory game birds.' He encourages members of the audience to join groups like the National Trust and the Royal Society for the Protection of Birds, which began in 1889 as The Plumage League to fight the trade in feathers for women's hats.

By the end of his Lectures, having shown both the wonders of rare animals and the problems they face, he leaves his audience with a positive outlook. As a reporter from the *Manchester Guardian* wrote on 10 January 1938, 'Much good, Mr Huxley showed, has been done; much more remains to do.' Eighty years later, that sentiment is still as true as ever.

CHAPTER 5

How Animals Move

Sir James Gray

1951

Moving through the world – whether on land, in the sea or air – is a defining feature of animal life. All sorts of creatures are invited in to creep, walk, swim and hop around the Lecture Theatre as Gray embarks on an energetic exploration of the way animals get themselves from place to place. And he reveals that no matter how animals move, they are all doing essentially the same thing: pushing backwards against the world around them.

As Gray walks into the Lecture Theatre, his table is covered in a selection of antique toy cars borrowed from the British Museum. He also shows the audience a cheetah. This particular big cat doesn't move as fast as it once did (it's a stuffed specimen from London's Natural History Museum) but, Gray explains, all living animals move in a similar way to motor cars: they need an engine, steering and brakes. He likens an animal's body to a car's chassis, and its legs to the back wheels (which do the pushing). Unlike cars, nature never uses a true wheel. Gray tells the audience that animals instead

Gray shows his audience a stuffed cheetah.

have rods or levers that move up and down or from side to side, but they can never make a complete revolution about an axis. Animals can't have wheels because all their moving parts are connected to blood vessels and nerves, which would get tangled and broken by constant twisting. Instead of trundling about on wheels, or flying with spinning propellers, animals, Gray explains, have evolved three distinct ways of moving.

'Let us look at nature's earliest types of animal life and try to see how movement was brought about,' Gray says. He shows a film of microscopic amoeba, single-celled creatures that are one-twentieth the size of a pinhead. They continually change shape and move by sticking out bulges from their bodies, known as pseudopods. This, Gray announces, is 'nature's first attempt to make a moving animal'. But they move very slowly, at around a foot (30 cm) per day.

The next film features creatures called ciliates, which look like transparent grains of rice that speed across the screen. They propel themselves by waggling thousands of fine hairs, called cilia (hence their name). This is the second way animals move, although only quite small ones. The forces exerted by these hairs are minute. Ciliates can travel perhaps 120 ft (36 m) a day.

All larger animals depend on muscles – the third way of moving – which are, according to Gray, 'the finest and most powerful of all nature's engines'. Muscles consist of bundles of fibres, each one instructed by nerve signals from the brain to contract and shorten, exerting a great force. The fibres, he says, change from 'a piece of perished rubber to a stretched steel spring'. As soon as the nerve impulses stop, the muscle fibres once again go limp. 'But they do not extend by themselves,' he says; 'they have to be pulled out by other muscles.' Nature harnesses the power of muscles in two ways. First, as in earthworms, muscles on both sides of the body contract at the same time, shortening the whole body like a concertina. By gripping the ground at the front end, this shortening pulls the animal forwards. It

Lecture programme (front cover).

then lengthens its body, and pushes forwards again, by contracting another set of muscles that form rings around its body, like squeezing a handful of putty. The second way muscles work is by contracting on one side of the body (or a limb) at a time, and bending the body from side to side.

To demonstrate a fundamental principle of movement – that in order to move forwards animals have to push backwards – Gray places a small bridge on his desk. The children hold their breath while he invites one of his earthworms to wriggle across. As it goes, a spring-loaded central arch shows that the worm pushes backwards with a force of between

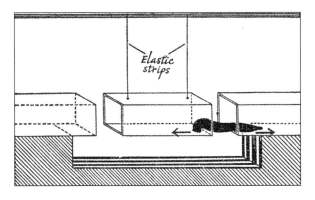

Diagram of an experimental bridge demonstrating the backwards force a leech creates as it moves forwards (similar to the worm bridge used in his Lectures).

Gray shows members of his audience a mechanical duck model.

two and eight grams (not bad for a small worm). 'As it completed its unnatural journey and dropped into the hands of its keeper, the worm was sympathetically applauded', wrote a *Times* reporter on 28 December 1951. Whether they're walking, creeping, jumping or swimming, all animals are doing essentially the same thing, and pushing back against their surroundings.

An aquarium tank is brought into the Lecture Theatre with a small turtle inside. It rows through the water using its flippers as paddles, again pushing backwards against its surroundings (in this case water). A mechanical model duck in another water tank shows a similar thing; it pushes against the water with its feet. Most fish swim through water in a different way, Gray explains, by bending their backbone to and fro. Powerful muscles swing a large tail from side to side, pressing on the water sideways and backwards, driving the fish forwards. Dolphins have a similar set-up, only their tails move up and down.

Gray also introduces a live eel to show how fish can only move if they have something to push against. First he lays it on a smooth table and the eel wriggles desperately but doesn't go anywhere. Then he places it on a table studded with pegs and it immediately braces itself against them, as it would push against water when it's swimming, and the eel crosses the table in a flash.

How fast can fish swim? The answer, Gray explains, is perhaps a little surprising. 'When we walk along the riverside and catch sight of a trout darting through the water,' he says, 'we get the impression of very high speed.' Anglers might guess that fish race by at 20 mph (32 kph). Fish can accelerate very quickly, which may lead to the false impression of great overall speed. When it's startled, Gray says, a trout accelerates at 140 feet per second (95 mph). Such powerful forces can only be sustained for a short time, but hopefully long enough to escape a predator. Over longer distances,

---

Sir James Gray (1891–1975)

Born in Wood Green, London, Gray studied natural sciences at King's College Cambridge. After serving in the First World War he returned to study in the Department of Zoology in Cambridge in 1919 where he eventually became professor. Early in his research, Gray helped establish the field of cytology, the study of cells, before switching to focus on animal locomotion. From 1943–7 he was the RI's Fullerian Professor of Physiology. He's remembered for Gray's Paradox, his theory that dolphin muscles aren't strong enough to overcome the drag forces in water when they swim very fast. In 2008 researchers proved him wrong, showing he had underestimated dolphins' muscle power by a factor of ten.

---

he estimates a ten-inch trout may swim at 4 or 5 mph; a large salmon might manage twice that speed. 'The really fast swimmers are the dolphins,' he says, which can swim as fast as 25 mph (40 kph).

Fishers are also prone to exaggerating the strength and weight of the fish on the end of their lines, and for good reason. Gray invites members of the

Pages of Gray's Lecture notes written on RI headed paper.

audience to try lifting blocks of different weights from the floor using fishing rods fixed with a spring balance to measure the force needed to lift the weight. To lift a one-pound weight (453 g), the volunteers have to sustain a pull of some 14 lbs (6.35 kg) on the rod. It means that a fish tends to feel heavier and stronger on the end of a line than it really is.

Next, the audience has a chance to explore the underwater world as Gray shows some previously unseen film footage, shot in the Mediterranean by 'frogmen' (otherwise known as scuba divers; it was only in the previous decade that Jacques-Yves Cousteau and Émile Gagnan had invented the aqualung). On screen, fish shoal in amazing numbers and varieties and a diver swims among them, even touching the tail of a shark that's nearly as long as he is – all sights the audience have probably never seen before (in the 1950s underwater filmmaking was just taking off; in 1956 Austrian biologist Hans Hass made a documentary *Diving to Adventure* and the same year saw the release of Cousteau's first major film *The Silent World*).

Like Thomson in his Christmas Lectures thirty-one years previously (see page 21), Gray introduces his audience to mudskippers, only this time they get to see not just preserved specimens but a living animal. It uses its two front fins and its tail to clamber around a tank of shallow water. 'From a creature of this type,' Gray says, 'nature has produced land living animals that propel themselves by means of their legs. She has done this without providing any new parts. She has drafted existing parts for new purposes.' Gray compares this to an engineer building a motor car from a submarine without taking it into dock or adding new components.

Feet probably first evolved to be webbed, Gray says, to spread the animals' weight and stop them from sinking into swampy, muddy ground at the water's edge. After that, animals evolved feet to suit other terrains. Gray brings in more live animals to examine their feet. There's a sloth in a cage, which uses big, hooked claws to hang upside-down on a branch. A gecko runs up a vertical sheet of glass and hangs upside-down (it's now known geckos achieve this amazing feat by using microscopic hairs, called setae, that cover their toes and form weak bonds with the surface they're climbing). And keepers from London Zoo bring in a small bear, called Henry, who shows his flat, human-like feet that allow him to stand upright on two hind legs.

Gray's next guest is a salamander, which he places on the projector to show the audience how four-legged animals commonly move their limbs in a diagonal sequence: right foreleg first, then left hind leg, left foreleg, and finally right hind leg. 'Three feet are always on the ground whilst the fourth is swinging forward in the air,' Gray says. In this way, he explains, the animal can stop at any point and it won't fall over because at least three of its feet will be planted on the ground, forming a stable triangle. But this only works at a sedate pace. To speed up, animals must sacrifice their ability to stop at any point without falling over. 'If an animal lifts a forefoot before the hind foot on the same side has come down,' he says, 'the body will roll towards that

Diagram of a galloping horse, showing how it never has more than two feet on the ground at the same time.

side. But as long as it does not wait too long, the roll will not be very great before it is checked by the hind foot coming down.' When a horse gallops it has only one hoof on the ground at a time and periods when it doesn't touch the ground at all.

Next Gray considers how animals repeatedly lift all their feet from the ground when they leap and jump. 'Most animals can jump to some extent,' he says. 'If we were asked to pick a team of jumpers we would probably pick a kangaroo, jerboa, frog, grasshopper or flea.' And fleas are the undisputed champions. They can jump 100 times their body length (while the world-record human high-jumper made it to just 2.45 m or 8 ft, a little over their own height).

All jumping animals, Gray points out, tend to share three characteristics: (1) they have very long hind legs, (2) their legs can tightly fold up under the body and quickly straighten out, and (3) they are usually quite small. The small body size is important because the height an animal can jump depends on the speed of take-off, and small muscles tend to act more quickly than big muscles. He shows the audience live locusts in action and a jerboa which gets loose and causes brief havoc as it hops around between the children's feet; it's a jumping rodent from deserts in Asia and Africa that looks like a mouse with enormous ears and great, long legs.

Finally, Gray fills the Lecture Theatre with paper aeroplanes as he introduces his audience to the details of animal flight. 'Flight depends on wings,' he says. The shape of wings is such that when held at a slight angle to the air current, air flows faster over the upper surface than the lower, which creates lift. At the same time, the moving air tends to drag the wing backwards. 'An animal can fly,' Gray says, 'provided it keeps its wings moving through the air in such a way as to ensure that the lift and drag forces . . . all combine to form an upward force equal to the animal's weight.'

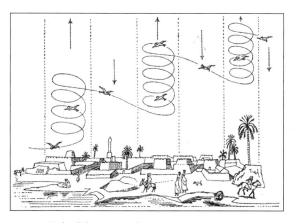

Birds gliding upwards on rising air currents in a tropical country.

As well as flapping flight, many birds are also expert gliders and often use rising air currents to keep themselves aloft. When horizontal winds hit an obstruction, like a building, a mountain or a cliff, the air current is deflected upwards; this is why seagulls glide along a line of cliffs and eagles glide on the windward side of mountains. A large, stuffed vulture is suspended on a wire above the audience as Gray tells them how in tropical countries, these birds spiral to great heights, riding on columns of hot air that rise up from the ground.

As Gray ends his story of how animals move, he tells the audience that he hopes they will now see that common animals around us all the time are in fact very complex and yet beautiful. 'I hope you will watch the animals for yourselves,' he concludes, 'for there is always something new to discover.'

An electric guest

Gray welcomes into the Lecture Theatre his colleague from Cambridge University, Hans Lissmann, who the previous year made a remarkable discovery. On a visit to the aquarium at London

Zoo, Lissmann noticed how the African knifefish on display swim forwards and backwards with equal ease, by rippling waves of the long fin along their back. He wondered how these fish manage to avoid bumping into things, given that they couldn't possibly see what's going on behind them, with eyes at the other end of their body. He guessed they have some kind of 'sixth sense'.

With the help of a live knifefish, Lissmann demonstrates how he discovered the weak electric pulses these fish send into the water. He lowers a probe into the tank that picks up the electric field and amplifies it so that the RI audience hears the buzzing on a loudspeaker. The fish, he explains, uses electricity in the same way bats use ultrasound to find their way around in the dark, by listening for echoes bouncing off objects around them. Lissmann was the first to work out that at least 100 species of fish living in muddy rivers and swamps in West Africa find their way around, locate food and perhaps even communicate with each other using electric pulses.

From the archives . . .

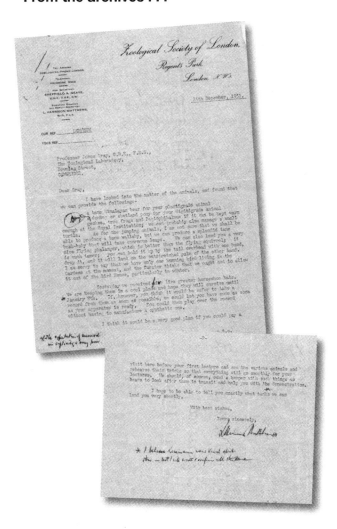

On 14 December 1951, an official from London Zoo wrote to James Gray outlining the live animals he could borrow for his Christmas Lectures. These included a tame Himalayan bear, a donkey, a gecko and a tree frog. A mudskipper was available, but only if Gray could ensure it stayed warm enough. 'We can probably also manage a small turtle', the letter states. It goes on: 'I am sorry to say that we have only one hummingbird living in the gardens at the moment, and the curator thinks that we ought not to allow it out of the Bird House, particularly in winter.'

CHAPTER 6

ANIMAL BEHAVIOUR

Desmond Morris
1964

✧

Throughout their lives, animals behave
in a huge variety of ways, to find food,
to survive threats and to make more
of themselves. With the help of live
guests from London Zoo, Desmond
Morris shows that by studying animals
in detail we can try to see the world
from their point of view and find out
what matters to them. Besides offering
fascinating insights into the inner
workings of the living world, Morris's
Lectures reveal that by understanding
other animals, we can learn more
about ourselves.

✧

'Nothing fascinates man quite so much as his own behaviour,' says Morris. 'From the depths of scandalous gossip to the heights of poetic expression, man comments on the ways of man.' It's the sphere of psychologists and anthropologists to analyse and understand human behaviour. 'But man is a complicated animal,' he points out, and it's not often clear why we do the things we do. We can, though, turn to other animals for help. 'There, with less difficulty,' he says, 'we can study the patterns of behaviour and learn some of the basic rules.' If we can learn to see the world from their perspective and understand their problems and the way they solve them then we can compare their way of life to our own.

Morris first explains what he means by behaviour. 'A stimulus from the outside world reaches an animal through one of its senses, such as the eyes, the ears or the nose. Messages from those sense organs are relayed

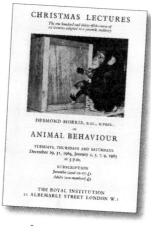

*Lecture programme
(front cover).*

87

along nerves to the brain.' Those messages are then combined with other, internal messages, such as memories from past experiences, and ultimately the brain responds by sending signals to different parts of the body – legs, arms, wings and jaws – instructing them to respond in certain ways. 'Behaviour is made up of a long sequence of responses, going on all the time from the animal's birth to its death,' Morris says.

A crucial facet of behaviour is the ability to learn, and some animals have evolved to be especially good

Morris introduces the audience to Butch, and some of the children reward the chimp with grapes.

at it. A young chimpanzee, called Fifi, is brought into the Lecture Theatre and Morris sets her a learning test. On the desk sits a large wooden cabinet with a door. While Fifi watches on, Morris opens the door, places a pile of grapes inside, then shuts the door, locks it and places the key on the desk. He then stands back while Fifi works out what she needs to do: before long she picks up the key, unlocks the door and tucks into the grapes. The audience give a round of applause at the young ape's dextrous and mental skills.

Morris goes on to explain the roles behaviour plays in different crucial stages of animal life. When animals come together to breed, all sorts of behaviours are involved, from males fighting over territories to the elaborate courtship rituals between mating partners (which we'll hear more about in David Attenborough's Lectures; see page 100). Once the young are born, parents find themselves extremely busy, with lots of new behaviours involved in rearing the next generation. Parental duties include many things from incubating and feeding, to defending young from enemies, and keeping them warm and clean.

It's also vital that parents recognize their own young and distinguish them from others of the same species. To demonstrate these abilities Morris brings in two female mice, one with black fur and the other white; each has a clutch of tiny offspring, with fur colours matching theirs. He puts the two nests on the desk, gently takes out the baby mice and places

---

Desmond Morris (b. 1928)

Born in Wiltshire, Morris studied zoology at the University of Birmingham, then did a PhD at Oxford University on the reproductive behaviour of stickleback fish. In 1956 he became head of the Television and Film Unit at London Zoo and from 1959–67 was the zoo's curator of mammals. He became famous for presenting the television shows *Zootime* and *Life in the Animal World*. In 1973 he returned to Oxford University to continue studying animal behaviour. He's written dozens of books on animals and the natural world including the bestseller *The Naked Ape* in 1967 and *The Human Zoo* in 1969. Morris is also a painter. He exhibited alongside Joan Miró in the 1950s, for a short while he was director of the Institute of Contemporary Arts in London, and he continues to exhibit his paintings.

them in the centre, mixing them up in a cluster of black and white infants. While he does this, Morris keeps the mothers shut inside their nests. He predicts that the two mothers will collect only their own babies and take them back to her nest, ignoring the others. A little nervously, unsure if the experiment will work, Morris lifts the entrances to the nests and the two female mice immediately scurry out and begin searching and sniffing across the desk until they spot the cluster of babies and dash over. To his great relief, the mice do exactly as he hoped: each mother picks up only her own babies and quickly carries them one by one back to the nest (as we'll see in Attenborough's Lectures, it could be the babies' high-pitched calling that alerts their mothers to their presence and identity). Morris tilts the two nests to the audience to show the black-furred mother surrounded by her black-furred babies and the white mice all in the other nest.

Young animals, themselves, must learn 'who they are', as Morris puts it. 'Some creatures know their own kind instinctively,' he says. 'Others, however, must be exposed to their own kind in infancy if they are to become socially integrated.' He explains that for some birds and mammals there's a brief window of time after they first open their eyes, when the young become 'imprinted' on their parents. This is one way animals become tame. If the first thing they see is a human they become 'humanized'. If they see both humans and members of their own species, they maintain allegiance towards both, like domestic dogs.

Various behaviours evolved that protect animals from daily threats to their survival. 'Cleaning behaviour may appear to be little more than a pleasant luxury for a wild creature, but this is far from the truth,' Morris says. 'The condition of an animal's coat may well be its passport to survival.' Damaged skin can lead to disease. Birds with ragged, dirty feathers may become chilled or overheated. 'A close study of the scratchings, shakings, wipings, preenings, groomings and rubbings of birds and mammals soon reveals that these are not occasional, random acts,' Morris says. In fact these are carefully organized behaviours, vital for keeping animals clean and healthy. To avoid extremes in temperature, many animals migrate to warmer climes in winter. 'Alternatively they prepare themselves for hibernation and sleep out the long, cold months in special dens or shelters,' he says. Animals also need to escape predators; some have evolved to run fast or fly away, others concentrate on hiding by burrowing and digging underground. 'Others have become strangely camouflaged,' he says, 'often with incredible complexity, so that they match perfectly the backgrounds on which they rest' (like the leaf insects we met in Julian Huxley's Lectures; see page 53).

'While they are avoiding becoming food themselves,' Morris says, 'all animals must try and find a meal. For many creatures this involves an almost non-stop search.' Animals obtain food in all sorts of ways, from gathering fruit, roots and leaves, to

Fifi assisting Morris with one of his demonstrations.

hunting and eating each other. At the most basic level, animals select food from their environment and need to distinguish what is good to eat, and what is not. Animals may then simply bite and swallow the food, or they may need to peel fruit skin, crack hard shells and nuts.

Morris welcomes another guest into the Lecture Theatre to demonstrate an important skill that ancient human ancestors developed for obtaining food, and which we still share today with some of our closest relatives. Butch the chimp is brought in as Morris explains to the audience the task ahead. He assembles

The coconut shy experiment, seen here a few years before the Lectures with children at London Zoo.

a coconut shy (a traditional fairground game) scaled down to the young chimp's size. On a wooden board stands a row of pegs with grapes balanced on each one, and above them hangs a long chain with a ball on the end. To get the grapes, Butch has to swing the ball and knock the pegs over. The audience watches on in delight as he has a few goes and gradually improves his aim until he manages to strike a peg and the grape rolls down into his waiting hand. It's not that early humans made fairground attractions like this, but learning to aim accurately at a target and throw weapons is a crucial hunting skill. It was, Morris explains, an evolutionary development that forged the character of the human species and a crucial step in the way people lived in small hunter-gatherer tribes.

 FROM DESMOND MORRIS . . .

'When I was asked to give the Royal Institution Christmas Lectures in 1964, on the subject of animal behaviour, I was determined to accept the challenge although at the same time I was alarmed at the prospect of having to present demonstrations using live animals in front of a live audience. The chances of something going wrong were much higher than with the more usual Christmas Lectures where experiments were performed in the scientific disciplines of physics, chemistry, engineering, and other non-biological subjects. As I knew from my television programme *Zootime*, transmitted weekly from the London Zoo, animals were highly unpredictable and I had visions of my demonstrations becoming somewhat chaotic. Why should I expect zoo animals to behave naturally in front of a large audience in unfamiliar surroundings in the centre of London?

'Despite my fears I started to work out various experiments and tests that I could try out using animals that would not be too disturbed by finding themselves in such an unnatural environment. I was lucky to have the assistance of keeper staff from the London Zoo who knew their animals intimately and were keen to help me in the same way that they did in my weekly television shows. When the time came I entered the lecture hall with more than the usual trepidation. It was packed with eager and intelligent children and their parents. To my great relief my demonstrations of patterns of animal behaviour all worked remarkably well.'

And whenever Morris's animals performed as he hoped they would, he made sure not to take the credit for himself: 'In the small room where a lecturer has to wait before he gives his talk at the Royal Institution there is a desk where you sit to think about what you're going to do. It is a quiet, peaceful room but as you try to relax there, you notice that there is a printed card on the desk with instructions to lecturers on what they may and may not do when speaking at the Royal Institution, especially when performing experiments. This card, which had been written by the great Faraday himself, was at pains to inform all speakers that they must never ever show any pride when an experiment goes well. I forget the exact wording, but it was very strict, insisting that tests and experiments and demonstrations should not be seen as something clever that the speaker had done, and that he must always maintain a modest demeanour. I followed this instruction carefully throughout all my demonstrations and always made it clear that it was the animals that were being clever and not me.'

From the archive . . .

In March 1964, Morris wrote to Sir Lawrence Bragg, thanking him for his invitation to give the

Christmas Lectures, and suggesting the title could be 'Animal Signals'. In September that year, Bragg writes to Morris, asking if he can make sure his Lectures can be 'put in words of one syllable' for his young audience to understand.

Dear Sir Lawrence,

Many thanks for your kind invitation to give the Christmas lectures. I am extremely flattered and I have every hope that I shall be able to give the complete series of six for you. May I please leave an absolutely final decision until May? I should know my future commitments more precisely by then. I am just about to go on leave but I will get in touch with you again as soon as I can give you a final answer. In the meantime, if May is too late for you, will you please drop me a line which my secretary will put before me as soon as I return from leave.

If I am able to give the lectures, the title will probably be "Animal Signals" and will give ample scope for the kind of demonstrations at which the Royal Institution excels.

Yours sincerely,

Mary Haynes.

Dictated by Dr. Desmond Morris
and signed in his absence.

---

16 September 1964

Dear Morris,

I was very pleased to get your notes this morning. If I may say so, I think the contents of each lecture are very well outlined and most attractive. I shall much look forward to this series. I always feel diffident about making any suggestions, but I think that as your audience and readers will be young people it perhaps might be a good plan to paraphrase or explain some of the technical words. The introduction is important as giving the first impression of the course. I will not try to make any specific suggestions, but the more it can be 'put in words of one syllable' the better it is for the **teenagers** and under. Again in Lecture 6 I think such words as 'socially-integrated', 'mal-imprinted' etc. - though it is clear from the context - could have a word or two of explanation.

These are very minor points and I feel most warmly that it is an **excellent** synopsis. I am so glad you are doing this series.

Yours very sincerely,

W. L. BRAGG

Dr Desmond Morris,
The Zoological Society,
Regent's Park,
N.W.1.

EYEWITNESS

Two members of the audience for Desmond Morris's Lectures were future RI Christmas Lecturer David Attenborough and his young son, Robert. 'My main sense is just that I enjoyed them, really,' says Robert, 'and they alerted me to the thrills of observing animals. It certainly would have made a big impression on me because I was interested in animal behaviour ever after.' Robert went on to become a biological anthropologist.

With a shared interest in zoology and television, the Attenborough family became good friends with Morris. At around the time of his Christmas Lectures, while he was London Zoo's curator of mammals, Robert remembers Morris giving his family a hoolock gibbon from Burma to look after. It had been brought to the zoo to partner a male gibbon, who attacked the new arrival – what was thought to be a female was in fact a young male. 'They had nowhere to put it, and it was ailing,' explains Robert, 'so Desmond allocated it to us. It lived with the family, in the house in its cage except when permitted out, for some years after that.'

CHAPTER 7

THE LANGUAGES OF ANIMALS

Sir David Attenborough

1973

We set off on an exploration around the planet in search of animal languages, in the consummate company of Sir David Attenborough. From tropical coral reefs to African savannah, we encounter a multitude of animals that send out noisy, scary, enticing, vibrant and smelly messages to each other. Attenborough brings a parade of live creatures into the Lecture Theatre and transports the audience to distant lands with spellbinding film footage, showing how learning the languages of animals reveals much about how the living world works.

Before this year's Lectures begin, the Theatre fills with the sounds of geese and cars honking, of beating drums, beeping Morse code and chirping grass-hoppers. These sounds, says Attenborough, all contain messages: 'They were languages.' It's not only sounds that animals use to communicate with each other. 'You can use vision, by gestures,' he says, flapping his arms as wings, 'or you can change your pattern in a sort of language, or indeed you could use smell.'

Attenborough begins with the way animals use languages to avoid getting into fights. 'It seems to be the rule that animals prefer to advertise threat, to say beware, rather than actually get down to blows,' he says. From under a black cloth he reveals a large, coiled snake. He taps his fingers on the glass case and we hear a rattling sound. It's a diamond-backed rattlesnake from the deserts of North America. Hollow scales on its tail warn intruders of its poisonous fangs.

*Lecture programme
(front cover).*

Sir David Attenborough (b. 1926)

Born in London and brought up in Leicester, Attenborough studied zoology at Cambridge University and in 1952 joined the BBC as a trainee producer. In 1954 he presented the *Zoo Quest* series, which saw him travelling the world, gathering exotic animals for London Zoo. From 1965–8 he was Controller of BBC2 and the BBC's Director of Programmes from 1969 until 1973 when he returned to filmmaking. In 1979 he wrote and presented the landmark series *Life on Earth*, followed by dozens more series exploring the many wonders of the natural world. He has thirty-two honorary degrees and at least fifteen species named after him including, in 2016, a bright blue dragonfly from Madagascar, *Acisoma attenboroughi*, that was awarded him in celebration of his ninetieth birthday.

The audience giggles as the next animal comes in. 'It's a skunk,' someone mumbles. The furry animal does have black and white stripes but, as Attenborough reveals, it's a zorilla – a type of polecat. Like a skunk, the zorilla's bold patterns warn not of a poisonous bite but of something very unpleasant. 'Beneath his tail he has got a gland, like a squirter,' says Attenborough, 'from which he squirts the most disgusting smelling fluid.' And the zorilla knows this as it struts across the plains of Africa. 'He walks very boldly, very upright with his hair and his black and white pattern well displayed, because he knows he's boss.'

Attenborough brings out another, much smaller animal that says 'beware' with its body language. The praying mantis raises its prickly forelegs like a boxer. 'He's only small,' says Attenborough, 'but if you were an ant, or a grasshopper or a fly, he would be just as frightening to us as an elephant is.'

Other animals pretend to be dangerous when in fact they are not. 'Sometimes animals actually bluff,' says Attenborough. A large green chameleon walks up Attenborough's arm. 'When they're very angry they go black with fury, and they hiss,' he says. 'Even though they can't do any harm to you or me,' he says, 'it can be very frightening.' In Madagascar, there are chameleons that terrify local people. 'One bite from it, they say, and you'll certainly die.'

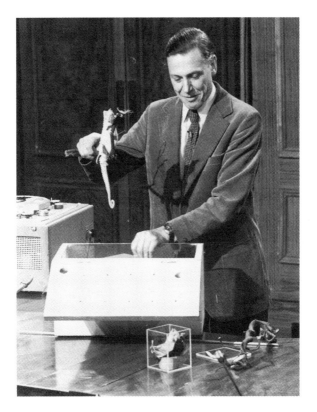

Attenborough shows his audience a live chameleon.

Gorillas put on fearsome displays, as the audience sees in a film of a huge male, beating his chest and roaring. Explorers discovered them in Africa's lowland forests, only 130 years before these Lectures, and assumed they were dangerous. 'So for years and years that poor beast was hunted,' Attenborough says. 'In fact, it is a very gentle, harmless creature.' (Attenborough found this out first-hand a few years later when he was famously filmed in Rwanda being groomed by a gorilla troop.)

The Lectures were broadcast live and Attenborough knew only too well how difficult it could be to get the animals to do things on demand. So, when the porcupine refuses to come out of its cage Attenborough chuckles and for a quiet moment tries to coax it out with a banana. Finally, it emerges into the small enclosure that's been erected around one entrance of the Lecture Theatre and children lean in to get a closer look. 'Now he's got a little alarmed. You see he's erected his quills,' Attenborough says. This animal isn't bluffing; porcupines will back into intruders, jabbing them with thousands of quills, up to 30 cm (1 ft) long and covered in barbs that easily puncture skin but are painful to remove. 'Can we get him back, do you think?' Attenborough says. Bill Coates, the RI's lecture assistant, lures the porcupine out of the Lecture Theatre with a trail of bananas.

Animals also have ways of warning members of their own species. 'Sometimes animals disagree amongst themselves,' Attenborough says. 'Sometimes they quarrel.' He shows film of male marine iguanas, wrestling on the shores of the Galapagos Islands. They don't bite, but nod and lock heads, then shove until one gives up and retreats.

Attenborough's next guest combines colour, gesture and smell in its threat displays. 'This is

17th January, 1974.

Dr. G. F. Claringbull,
Director,
British Museum (Natural History),
Cromwell Road,
London, S.W.7.

Royal Institution Christmas Lectures

Now that the Christmas Lectures are over, I am writing to thank you for so kindly agreeing to loan various specimens from the British Museum for demonstration purposes in the Christmas Lectures. These added greatly to the interest of the lectures and we do appreciate your help.

The lectures were very successful and you will probably be interested to know that the viewing audience topped the million mark on four occasions, in addition to the very enthusiastic 'live' audience here in the Royal Institution.

With kind regards,

A thank-you letter from RI Director Sir George Porter to the British Museum, sent after the Lectures.

Tammy, from Bristol Zoo, and he's a ring-tailed lemur. Aren't you?' he coos, clearly enchanted by the cuddly creature in his arms. 'Would you like a bit of grape?' Tammy chirps and tucks in, to the obvious delight of the audience. Attenborough tries to point out the scent glands on the lemur's wrists, not an easy task with Tammy busily eating. 'If Tammy wants to fight, he will wipe his lovely black and white striped tail through those glands,' Attenborough says. 'Then he will engage in what's called a stink fight.' The audience giggles at this idea. 'He doesn't actually throw stink bombs, but what he does do is to lift up his tail – I beg your pardon . . .' (Tammy squeaks). Attenborough continues, '. . . and waves it over his back so that he creates a current of air which blows this nasty smell at his enemies.' Reluctantly, Attenborough gives Tammy back to his handler and then shows his young audience a film of lemurs stink fighting in Madagascar.

The haunting calls of another lemur species echo around the Lecture Theatre, as Attenborough considers how animals use language to defend territories. It's an indri, the largest living lemur species. He used this recording to help film lemurs in the dense forests of eastern Madagascar. On the screen

Attenborough introduces the audience to Tammy,
a ring-tailed lemur.

the audience sees nothing but trees until eventually a large, black and white animal appears clinging to a tree, like a giant teddy bear. 'He seemed to be fascinated by this recording,' Attenborough says. 'I suspect that he had never imagined that another rather tinny indri could have invaded his territory.' The indri calls back with a loud honking noise. Attenborough explains the meanings of the two territorial calls. 'The first was that wailing call that I played which says, "This is my land, this is my territory."' The honking sound says 'Get out!' to an intruder.

Having established a territory, the next step is for an animal, usually the male, to attract a partner. Attenborough introduces various ways animals say 'be mine'. Film footage of male frigate birds show the inflatable red sacs on their throats, which display to passing females. The audience then hear the nesting calls of blackbirds and Attenborough asks the audience what the males might be saying. 'Where the bird is . . . what kind of bird . . . how good a singer it is.' All good answers. He also points out that every blackbird has a slightly different song. 'It is absolutely possible that female blackbirds can recognize their own mates. Not only that, but blackbirds in this country have different accents.' The audience chuckles at this. 'It is possible for a blackbird to sing with a Yorkshire accent.'

The audience immediately recognize the call of the next animal. 'What a well-informed lot you are,' Attenborough says, as several children shout out 'Whale!' Only a few years before these Lectures, in 1967, a scientist called Roger Payne confirmed humpback whales sing songs. 'And we don't know whether that is a mating song,' says Attenborough, and 'whether in fact, this huge fifty-foot-long animal is lying down in the depths of the ocean singing this song, maybe calling to a female a hundred miles away ... One of the tragedies, of course, is that we may never know what that song is ... The humpback whale is becoming increasingly rare and is getting closer and closer to extinction, because we human beings kill humpback whales in order to make margarine and soap, which you may well think is a crime and a scandal.' (Five years after these Lectures, the 'Save the Whale' campaign brought a halt to commercial whaling; Payne's recordings of whale songs played a key part in the campaign's success, helping to swing public opinion in favour of whales by revealing their complex, beautiful songs. Many whale populations are now increasing and researchers are still working on deciphering their songs; only males sing, so it's likely these are some kind of mating call.)

Attenborough with members of his audience, who demonstrate their ability to waggle their ears, as other primates do, to communicate with each other.

Once courtship and mating is over, language continues to be important in the lives of animals. Attenborough shows the audience a pile of huge eggs from a South American ostrich called the rhea. 'Or at least,' he admits, 'since we're in the Royal Institution, the home of scientific accuracy, I'd better be a little more truthful than that.' Only some are genuine eggs, the rest are plastic replicas making up a total of fifty-four – which just happens to be the average number of eggs that a male rhea watches over and incubates (the eggs come from multiple females). Attenborough explains how important it is for the male rhea to hatch all his eggs simultaneously. 'It's going to be a frightful business if, say, twenty hatch today and start scampering around. And then he has to go and chase after those and abandon his nest.' In fact, the unborn chicks talk to each other through their shells and coordinate their hatching times. Attenborough demonstrates this with a set of much smaller quail eggs in an incubator, sitting on special microphones, which pick up the chicks' tapping sounds. He describes the recent experiments of Margaret Vince at Cambridge University, who showed that one chick's tapping tells others to speed up and hatch.

When they're born, young animals then communicate with their parents. Attenborough attempts to show that baby mice emit high-pitched squeaks that their mother responds to. In a Perspex box, he moves the pink, squirming babies from their nest to a clear chamber through which the mother should hear their calls, but she shows little interest in finding them. He returns the babies to their nest, only for the mother immediately to investigate the now empty chamber, and the audience roars with laughter. 'You know I never liked mice,' says Attenborough, scratching his head, 'and I like this one least of all.'

As well as communications between parents and offspring, some animals learn the languages of other species, including the language of light. 'There's a great deal we do not yet know about luminescence, about the way in which animals produce light,' he says. 'But we know approximately how it goes.' The Lecture Theatre lights dim and in a glass dish Attenborough mixes two liquids, which briefly glow blue. 'I think that is so marvellous,' he says, grinning. 'Let's do it on a bigger scale.' He brings out a large, round glass flask and rubs his hands in anticipation. 'How about that?' he declares as the flask illuminates in a bright swirl. Animals do a similar thing, mixing light-producing chemicals, and Attenborough has a model of one of them – a brown beetle, commonly known as the firefly. A light on its bottom flashes when Attenborough presses a button. Male and female fireflies find each other in the dark with specific flashing patterns. A male of one species emits two flashes separated by a longer pause; the female responds with a single flash. The pair come together and, Attenborough explains, 'they mate and everything's splendid' – unless, that is, a female firefly from another species gets involved. She's learnt to eavesdrop on these signals and emits the same single flash. When a male comes near, instead of mating with him she eats him. 'Only when she gets a completely different signal from the males of her own kind will she mate.'

Many animal languages are relatively simple but in some cases the messages can be quite complex. One of the most complicated is the honeybees' waggle dance. In the same year of Attenborough's Lectures, 1973, Austrian scientist Karl von Frisch was awarded a Nobel Prize for his work in discovering this sophisticated form of insect communication. He worked out that when a foraging bee finds flowers laden with pollen, it returns to the hive and tells other bees where to find them by dancing in a particular way. To give the audience an idea of what this looks like Attenborough sets up a large model of a dancing bee. Mr Coates wheels in a trolley with a large model honeybee on a track (which looks like it came from a model railway). Attenborough flips a switch and the bee begins to trundle around. With a bit of persuasion, the model bee performs loops in a figure of eight and in the middle section wiggles from side to side. 'There we are,' says a relieved Attenborough.

He explains how the direction of the bee's dance indicates to other bees the direction of nectar-laden flowers. He flips another switch and his model bee moves faster around the track, and wiggles its bottom more vigorously in the middle of the figure of eight. In a real hive, the faster a bee dances, the closer the flowers are located. Other details of the bees' waggle dance, which are too complex for Attenborough's model, indicate the flowers' angle to the sun. 'And so,' Attenborough says, 'you have an example of the most complicated piece of information passing in the insect kingdom.' He admits that the waggle dance is highly unusual among the insects. 'It is very exceptional and very extraordinary.'

From the archives . . .

On 5 January 1973, David Attenborough wrote to thank the RI's Director, Sir George Porter, for inviting him to give the Christmas Lectures. Attenborough suggested he could talk about animal colours and patterns, a topic that eventually expanded into the languages of animals.

5.1.73

Dear Sir George,

Thank you so much for your letter. I am very flattered indeed to be invited to give your Christmas lectures. In principle I would like to accept very much indeed. My only hesitation stems from doubt about the subject — what it should be and whether I can do it adequately. My initial reaction is that, needing to have

Something that is visually interesting and susceptible of experimental demonstration, we might select the meaning of the shapes, colours + patterns of animals, with individual lectures on camouflage, aggression, mimicry, courtship and so on.

It would not, of course, be very new, but I am not sure how important you regard that.

May I think about it for a few days + then come + see you + discuss it?

Thank you again for inviting me. I take it as a great compliment.

Yours sincerely,

David Attenborough

Attenborough had misgivings beforehand about giving live Lectures involving animals, but these doubts proved to be unfounded and the series was a huge success. Porter wrote to Attenborough after the Lectures to thank him.

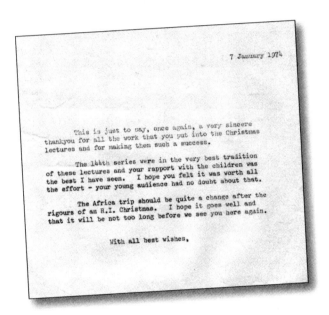

7 January 1974

This is just to say, once again, a very sincere thankyou for all the work that you put into the Christmas lectures and for making them such a success.

The 144th series were in the very best tradition of these lectures and your rapport with the children was the best I have seen. I hope you felt it was worth all the effort - your young audience had no doubt about that.

The Africa trip should be quite a change after the rigours of an R.I. Christmas. I hope it goes well and that it will be not too long before we see you here again.

With all best wishes,

Attenborough replied the following week to Porter, revealing that since the Lectures he'd had a dream that the recording of the Lectures had been faulty, and that he'd had to do them all over again!

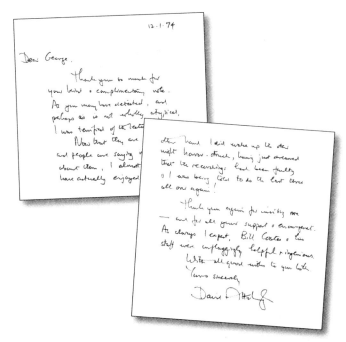

The head teacher from a primary school in Luton also wrote to the RI in January 1974: 'I was recently most impressed and at times disarmed by the David Attenborough Christmas Lectures . . . I only wish more of the children at my school had watched these lectures.' Also in January, a letter from Marie F. Harmer, from Birmingham, appeared in the *Radio Times*: 'He has a rare talent for keeping a vital interest in the subject, at the same time being human and humorous.'

CHAPTER 8

GROWING UP IN THE UNIVERSE

Richard Dawkins

1991

✧

To see how the great wonders of life evolved, Richard Dawkins takes us on a hike up the slopes of 'Mount Improbable'. It's the first time the Christmas Lectures have dealt directly with the topic of evolution and Dawkins, as ever, doesn't shy away from stirring debate. He warns that just because plants and animals can appear to be designed – it doesn't mean they were created by an intelligent higher being. And we find out how the universe, and life on our planet, grows up by gradual degrees.

✧

'Where does life come from? What is it? Why are we here? What are we for? What is the meaning of life?' Dawkins throws out these big questions to the audience early on in his first Lecture. 'There's a conventional wisdom which says science has nothing to say about such questions.' Dawkins, of course, begs to differ.

Throughout his Lectures, Dawkins criticizes what he calls superstitious and parochial views of the universe, including the origins of life on earth. In particular, he challenges the view that living things must be the work of a divine, conscious creator, or an intelligent designer. 'Growing up in the universe partly means evolving from simple to complicated, inefficient to efficient, brainless to brainy,' he says. We also grow up, he says, by gaining 'a proper scientific understanding of the universe, based upon evidence, public argument, rather than authority or tradition or private revelation'.

Lecture programme (front cover).

He demonstrates his own trust in science with the help of a heavy metal cannonball, the size of a large grapefruit, which hangs on a rope from the ceiling. Standing against the far wall, Dawkins holds the ball to his forehead and prepares to let it swing like a pendulum across the room and back again. 'All my instincts are going to tell me to run for it, but I have enough faith in the scientific method to know that it's going to stop just about an inch short, perhaps less, of my head,' he says (air resistance will slow the cannonball down and it won't return to the exact same spot). 'So here goes.' He lets go and a few seconds later the cannonball comes back, almost touching the end of his nose, and the audience erupts into applause.

Before diving into the details of evolution, Dawkins reminds his young audience of just how improbable and magnificent our living planet is. He asks them to imagine they're in suspended animation on a spaceship hurtling through space, searching for habitable planets. The ship is almost unthinkably lucky, he says, and lands on a planet capable of sustaining life, filled with colourful plants and animals. 'You would surely walk around in a trance,' he says, 'unable to believe the wonders that met your eyes and ears.'

People often don't open their eyes to the great wonders of life on planet earth because of what Dawkins calls the 'anaesthetic of familiarity'. He suggests a few ways we can try to awaken our fascination. He shows a fantastical picture of Jupiter, imagined by cosmologist and RI Christmas Lecturer Carl Sagan, inhabited by bizarre creatures he called Jovians. Dawkins invents another word – the Bijovians – for real animals that are so odd they almost could be from another planet. He shows a film of a cuttlefish with complex patterns flickering over its body and a model of a deep-sea anglerfish with a glowing lure dangling in front of its toothy mouth. 'A very weird Bijovian creature,' he says.

We can also shake off our indifference by going on a journey into the world of the very small. In the Lecture Theatre stands a Scanning Electron Microscope (SEM), which instead of light uses electrons to produce detailed images of tiny things. He invites Louise from the audience to use a joystick to navigate around the SEM's glowing green screen and what looks at first like a tropical rainforest. She zooms out, revealing the head of a mosquito with feathery antennae. Next we move on to consider the nature of the electron microscope itself. This complex device was clearly designed by people. Dawkins compares it with *designoid* objects, things that look as though they were designed. A young volunteer, Andrew, walks in with an example of a designoid object draped around his neck. It's a Boa constrictor, which looks like it's been designed for sneaking up and catching prey. 'What the Boa constrictor is best at,' Dawkins says, 'is throttling its prey.' The audience laughs and Andrew nervously goes off to be untangled from the snake.

Andrew gets acquainted with a Boa constrictor, an example of a designoid *object.*

The audience sees more examples of designoid objects. There's a pitcher plant, which appears well designed for catching insects. Camouflaged animals seem designed to look like other things; there's a picture of a leafy sea dragon that resembles a frond of seaweed ('one of my favourite designoid objects,' Dawkins says) and butterflies with wings like dried leaves. The question of how all these designoid objects came to be was answered surprisingly recently in the middle of the nineteenth century. This was, he says, 'one of the greatest discoveries of all time, made by one of the greatest scientists of all time – Charles Robert Darwin'.

Dawkins picks up a first edition copy of *On the Origin of Species*, which lays out Darwin's theory of evolution by natural selection. The book begins by discussing a different process called artificial selection, or selective breeding. To show some outcomes of artificial selection, Dawkins welcomes in a Great Dane, a German Shepherd and a Chihuahua. They bark and scuffle while Dawkins explains how they were all bred from the same wolf ancestors, which looked similar to the German Shepherd. Over generations, dog breeders selected puppies with particular desired characteristics – for example, the bigger or smaller ones – and bred them together, producing bigger or smaller dogs and ultimately something like a Great Dane or a Chihuahua. He also shows a selection of fancy pigeons and a range of vegetables – cauliflower, broccoli, kohlrabi – which were all selectively bred from wild ancestors.

In natural selection, instead of people doing the selecting, nature does. Dawkins demonstrates this with a computer program that evolves spider webs. The computer (taking on the role of nature) selects the webs that would catch the most flies and 'breeds' from them through repeated generations, each time adding minor changes to the webs' shapes. Fast-forwarding the output, Dawkins shows how the webs begin with bare spokes and gradually become filled in with spiralling, fly-catching threads. He explains how a similar thing happens in nature: spiders that build successful webs and catch lots of flies are more likely to breed and pass on the genes for building those webs. 'As the generations go by, webs get better and better,' he says. 'Exactly the same principle works for the body of every living creature. The Darwinian view is that designoid objects are not designed at all; they have evolved by natural selection.'

Richard Dawkins (b. 1941)

Born in Nairobi, Dawkins studied zoology at Balliol College, Oxford, and stayed on to complete his doctorate. For two years he taught at the University of California, Berkeley, before returning to Oxford in 1970. In 1976 he published his bestselling book *The Selfish Gene*, which explored a gene-centric view of evolution. His later books include *The Blind Watchmaker*, which won the 1987 Royal Society of Literature Award, *Climbing Mount Improbable* in 1996 and *The God Delusion* in 2006. From 1995 to 2008 he was the Simonyi Professor for the Public Understanding of Science at the University of Oxford. He's a Fellow of the Royal Society.

A challenge for Darwin's theory of natural selection is to explain how designoid objects evolve without needing huge amounts of luck. Dawkins quotes the eminent astronomer, Fred Hoyle, who once likened the chance of any complex living creature springing into existence by sheer luck alone to a hurricane sweeping through a junkyard and spontaneously assembling a Boeing 747 aeroplane. Dawkins goes on to explain that luck is spread out by natural selection. 'We don't have to get all our luck in one ridiculously large dollop,' he says. Instead it can come in dribs and drabs, with 'each drib being allowed to count before the next drab'.

He explains this further with a pair of computerized monkeys – called Hoyle and Darwin – which are tasked with typing a line of Shakespeare: 'More giddy in my desires than a monkey' (from the play *As You Like It*). Hoyle types at random. After every line the computer checks to see if it's written the complete phrase. 'If it does,' Dawkins explains, 'it will be the most improbable coincidence in the history of the world, and I solemnly promise to eat my hat.' He puts on a bowler hat.

The other monkey, Darwin, does the same thing, only with a crucial difference. When he types a random phrase the computer makes fifty copies with small mutations, here and there changing a few letters. The computer then picks the phrase most closely matching the target, however slightly, and uses it to make another fifty copies with mutations. It then repeats this process over and over. Typed phrases from the two virtual monkeys flicker on the Lecture Theatre's screen. Darwin's line quickly begins to look familiar. It takes a few seconds, and just 158 generations, for Darwin to write the complete phrase, while Hoyle is still typing gobbledygook. 'Things like eyes and 747s that couldn't possibly spring into existence in a single lucky shake of a dice,' Dawkins says, 'can come into existence if the luck is smeared out in many tiny steps and is accumulated.'

Next, Dawkins shows the audience a model of a mountain he calls Mount Improbable. The top of

Dawkins with his model of Mount Improbable.

the mountain is where we would find designoid objects, like an eye or an elephant. At the bottom are distant ancestors that are less well suited to their environment. Between the two is a vertical cliff. 'Jumping from the bottom of the cliff to the top corresponds to assembling a 747 by means of a hurricane,' he says, 'or getting a complete eye in a single mutation. It can't be done.'

But that's not the only way to reach the top of Mount Improbable. He spins the model around and reveals on the other side a series of sloping paths towards the peak – what he calls the ramp of evolution. Individual animals don't climb Mount Improbable, he says, but lineages and groups of one species do, over long stretches of evolutionary time. As an example he welcomes an eagle and an owl into the Lecture Theatre to show off their elegant wings. These expert fliers would perch at the top of Dawkins's metaphorical mountain, but how did they get there? If their wings gradually evolved, can half a wing be any use?

Creatures made from Christmas baubles demonstrate that it can. They have googly eyes, pipe-cleaners for legs and some have a skirt-like flap of card, representing simple wing stubs – the beginnings of wings. Dawkins releases one pair from a low height and they both land intact. A second pair falls from a greater height. This time the creature with wing stubs lands safely while the wingless creature is smashed to pieces. 'Controlled gliding has evolved many times,' Dawkins says, showing film clips of flying squirrels, snakes and lizards gliding between trees. These animals all have, as he puts it, half or quarter wings. As they evolved to survive greater falls and glide further, the wings of the birds' ancestors (and bats and insects) gradually became better at gliding and flying.

In his riskiest demonstration, Dawkins introduces the bombardier beetles. When they're attacked, these insects mix chemicals in their bodies and spray the resulting explosive, boiling brew at their enemies. Creationists, he says, often use these beetles as a case against evolution. They argue that only a divine creator could design a beetle that doesn't blow itself up; it can't have gradually evolved. To the obvious delight of the audience, Dawkins announces he's going to take on the creationists' claim and try being one of these beetles. He dons a tin soldier's helmet and holds up two beakers containing the chemicals the beetles use: hydrogen peroxide and hydroquinone. 'Anyone who wants to, can leave the room,' he says. But everyone stays glued to their seats, as Dawkins mixes the two liquids together. 'It's not even warm,' he says, to the disappointed audience.

The beetles' secret is a third chemical, a catalyst, that speeds up the reaction – and Dawkins has some of that too. He adds a sprinkle to a weak mixture of the chemicals, which fizz gently. 'That might slightly deter a predator but it would not be particularly dangerous to the beetle,' he says.

Next he tries a slightly stronger mixture, which gently steams and bubbles. 'That's distinctly warm,' he says. Finally, he tries a concentrated mixture. It gets boiling hot and sends up a dramatic whoosh of steam. Each of these stages represents the gradual evolution of the beetles' defensive tactics. Over generations, the beetles slowly climbed up Mount Improbable until they were spraying their lethal mixture from the top.

Next, Dawkins invites his audience to step into an imaginary ultraviolet garden. He shows film of flowers shot using a UV-sensitive camera, revealing patterns normally invisible to the human eye. And

Dawkins during his demonstration of the exploding defences of the bombardier beetle.

he poses the question: 'What are flowers good for? A human-centric answer might be that flowers are there to help bees make honey for us to eat. 'We need to find an entirely new view of the world,' Dawkins says. 'We need to try to see things through the eyes of other creatures instead of all the time through our own self-interested eyes.' From a bee's perspective, flowers are there to provide them with food (nectar and pollen). From the flowers' perspective, bees are working for them, moving their pollen to other flowers and ensuring they don't become inbred. The flowers' bright colours – including in UV, which insects can see – are adverts aimed at those pollinators. 'Flowers use bees, and bees use flowers,' Dawkins explains.

From yet another perspective, he introduces the idea of a 'total self-copying program'. It's an imaginary computer program that spreads copies of itself, like a virus. But instead of using existing computers to do so, it contains extra instructions to gather raw materials and build more computers. He imagines an autonomous robot with arms and legs and an on-board brain that constructs copies of itself from scratch. Such a machine has never been built.

Or has it?

A large chameleon walks ponderously along a twig on Dawkins's desk. This and all other living creatures, he says, are ruled by their own 'total self-copying program'. It's not written as a computer program but in the language of DNA. 'We are machines built by DNA whose purpose is to make more copies of the same DNA.' From this perspective, the purpose of flowers is to distribute instructions for making flowers. Likewise, bees are for making bees, birds for making birds, and so on.

In comes a parrot. 'A macaw's coloured feathers are for spreading copies of instructions for making more coloured feathers,' Dawkins says. The bright colours attract mates, so the genes for colourful feathers are passed on to future generations.

Plants don't have wings, he quite rightly points out, but they borrow the wings of bees and birds and use them to move their pollen around. So, from the point of view of the plant's DNA, the bees' wings might just as well be the plant's wings. They transport plant's genes, just as a macaw's wings carry a macaw's genes about. 'Now that really is a different way of looking at things,' he says, 'a strange and unfamiliar way.' And yet, he adds, it's a way of looking which 'matches the strange other-worldliness of the ultraviolet garden'.

A special guest

In his Lecture on the ultraviolet garden, Dawkins reveals that one of his favourite books is *The Hitchhiker's Guide to the Galaxy* by Douglas Adams. He asks for a volunteer to read from another book in the series (*The Restaurant at the End of the*

Universe). Adams himself is sitting in the audience and eagerly puts up his hand to read an extract from his book. In the 1980s, Dawkins wrote a fan letter to Adams and the pair remained good friends until Adams's untimely death in 2001.

Adams reads an extract that plays on the anthropocentric notion of animals existing purely for the good of people. Similar to Dawkins's ultraviolet garden, and the idea that bees are there to make honey for us to spread on toast, Adams imagines an animal that actually wants people to eat it. At 6 foot 5 (1.96 m) Adams towers over the children in the audience, who giggle and laugh along as he reads:

A large dairy animal approached Zaphod Beeblebrox's table, a meaty bovine quadruped with watery eyes, small horns and an ingratiating smile on its lips.

'Good evening,' it lowed and sat back heavily on its haunches, 'I am the main dish of the day. May I interest you in parts of my body?'

Its gaze was met by looks of startled bewilderment from Arthur and naked hunger from Zaphod Beeblebrox.

'Something off the shoulder perhaps?' suggested the animal, 'Braised in a white wine sauce?'

'Er, your shoulder?' said Arthur in a horrified whisper.

'But naturally my shoulder, sir,' mooed the animal contentedly, 'nobody else's is mine to offer.'

'You mean this animal actually wants us to eat it?' exclaimed Arthur. 'That's absolutely horrible. It's the most revolting thing I've ever heard.'

'What's the problem, Earthman?' said Zaphod.

'I just don't want to eat an animal that's standing here inviting me to,' said Arthur, 'it's heartless.'

'Better than eating an animal that doesn't want to be eaten,' said Zaphod.

'But that's not the point,' protested Arthur. Then he thought about it for a moment. 'All right,' he said, 'maybe it is the point. I don't care, I don't want to think about it now. I'll just . . . I'll just have a green salad.'

From the press

Dawkins's Lectures stirred public debate on the matters of evolution and creationism. In a letter to the *Radio Times* (2 January 1992), Mrs J. Firth from Bradford wrote to say that she found the Lectures entertaining and thought provoking but questioned their coinciding with the Christian festival of Christmas. 'May I suggest that the series ought to have coincided with a pagan festival, such as Midsummer's Day, and be called the *Royal Institution Solstice Lectures*?' Dawkins replied: 'I am delighted that Mrs Firth, like many other people on both sides of the religious divide, enjoyed the lectures ... The tradition of these scientific lectures at Christmas time goes back a long way and it would be a pity to break it. They are not obviously more un-Christian than reindeer, presents and mistletoe.'

 FROM RICHARD DAWKINS . . .

In the second instalment of his memoirs *Brief Candle in the Darkness* from 2015, Dawkins reflects on his time at the RI. He writes:

In the spring of 1991, the telephone rang and a pleasant voice with a gentle Welsh lilt announced: 'This is John Thomas.' Sir John Meurig Thomas FRS, a scientist of distinction and the Director of the Royal Institution (RI) in London, was ringing to invite me to give the Royal Institution Christmas Lectures for Children, and I went hot and cold as he did so. The warm flush of pleasure at the honour was swiftly followed by a cold wave of trepidation. I immediately knew I would not be able to refuse the commission, and yet I lacked the confidence that I could do it justice.

Of course, he did accept and went on to discover just how influential the title of RI Christmas Lecturer would be:

One agreeable and unanticipated feature of the Christmas Lectures was that the very name was a golden key to unlock goodwill whichever way I turned. 'You want to borrow an eagle? Well, that's difficult, I honestly don't see how we can realistically, I mean, do you seriously expect ... Oh, you're giving the Royal Institution Christmas Lectures? Why didn't you say so before? Of course. How many eagles do you need?'

From the archive . . .

Among the archived material for these Lectures is a copy of the invitation sent out by the RI for a tea party that was held after Dawkins's final Lecture.

The Director and Mrs Day

invite

....

to a tea party to be held in the Director's Flat on
Saturday the 21st of December,
immediately after the second of this year's Christmas Lectures.

RSVP
by December 13th

Tel: 071 493 2710

Also stored are some of Dawkins's Lecture notes, with his handwritten annotations.

5

Can you imagine how it would feel, if you woke up, perhaps after a hundred million years of sleep, and found yourself on such a world. A whole new world, a beautiful planet of greens and blues and sparkling streams and white waterfalls, a world filled with tens of thousands of species of strange coloured creatures, darting, swimming and flying. You would surely bless your luck in arriving on such a rare world, and walk around in a daze, a trance, unable to believe the wonders that met your eyes and ears.

Fantasy painting

Well, this will almost certainly never happen to us.

And yet, in a way, that is just what *has* happened to all of us. We *have* woken up after hundreds of millions of years asleep. We *do* suddenly and mysteriously find ourselves on a world that sustains our kind of life, a beautiful world filled with tens of thousands of species. Admittedly we didn't arrive by spaceship, we arrived by being born. But the wonder of the planet, the dazzling surprise of it, that ought to be the same whether we arrive by spaceship or by birth canal!

We are amazingly lucky to be here. We are lucky that it is we who are here, not the countless millions of alternative people who would have been here if their great great grandparents had met instead of ours. There is also the lucky fact that the planet on which we have woken up is a very rare planet, one filled with life in the midst of a vast dust-cloud of dead, desert planets. Once again, it is obvious that our planet *has* to be one of the rare life-bearing planets because otherwise we wouldn't be thinking about the matter. But, again, this should not stop us marvelling at our privilege of being able to witness the remarkable sights and sounds of a planet laden with life. And we must not waste the privilege.

Here, it seems to me, lies the best answer to those narrow-minded people who are always asking what is the use of science. Many of you will have heard of Michael Faraday, one of British science's great heroes and the founder of these Christmas Lectures.

Portrait or bust of Michael Faraday

He was once asked by Sir Robert Peel - probably in this very room - what was the use of science. "Sir," Faraday replied, "Of what use is a baby?"

Have baby brought on, and hold it (if quiet), while talking

I always used to think that what Faraday meant by this was that a baby might be no use for anything at present, but it had great potential for the future. I now like to think that he might also have meant

CHAPTER 9

THE HISTORY IN OUR BONES

Simon Conway Morris

1996

Far from being just a bunch of dead bones, fossils are the key to finding out what life used to be like on earth, long before humans came along. We explore the ancient world as Conway Morris brings fossils to life and shows us how to read the traces and remains left behind by long-gone creatures. Along the way we'll meet bizarre animals like none alive today, peer into the lives of the most famous extinct animals – the dinosaurs – and trace our own ancestry through the rocks, back 3 billion years to the very beginning of life on earth.

'In the crust of the earth there are millions upon millions of fossils,' says Conway Morris. A good number of them surround him in the Lecture Theatre. He selects a huge fossilized tooth from a hippopotamus that lived 150,000 years ago on the site now occupied by Trafalgar Square. The tooth tells us that London was once hot and tropical. Then he picks up the huge skull of a woolly rhinoceros along with its magnificent horn that's longer than his outstretched arm. This animal also strode around ancient London, around 60,000 years ago, when Europe was in the grip of an ice age. 'Scientists are

Conway Morris shows his audience a huge T-Rex skull.

detectives,' Conway Morris says. 'We're dealing all the time with clues.' Fossils are the main clues he searches for, to find out what the ancient world used to be like.

To demonstrate the process of fossil hunting, Conway Morris brings out a rock from Lyme Regis on Britain's southern coast. Whacking it with a hammer and chisel, he neatly splits it in two. 'Well, that's marvellous for lots of reasons,' he says, as he shows the audience a cluster of shining spirals. 'These are fossil ammonites.' This is the first time the remains of these ancient molluscs, relatives of octopuses and snails, have seen the light of day since they died millions of years ago.

Conway Morris explains how fossils form when animal bones and shells, like those of the ammonites, are buried in sediments and gradually become replaced by minerals, turning them to rock. Fossils form in other ways too. He holds up a chunk of amber with a 40-million-year-old spider trapped inside. Some fossils are preserved in ice, most famously the woolly mammoths. Sadly, Conway Morris doesn't bring a whole frozen mammoth into the Lecture Theatre, but he does have a tuft of fur. (There's little chance of finding intact dinosaur

DNA inside mosquitoes trapped in amber, like they did in *Jurassic Park*, but there are now plans to recreate mammoths from DNA extracted from ice-preserved bodies.)

As well as their bodies, animals leave other things behind that tell us about the ancient world. 'Here is an animal going for a walk,' says Conway Morris, pointing out a trackway winding across a slab of limestone and, at the end, a fossilized horseshoe crab (similar to the living horseshoe crab Julian Huxley showed in his Lectures; see page 53). He describes the ancient animal being caught in a storm and swept into a lagoon, where it became intoxicated by poisonous water; it stumbled in circles across soft mud, leaving its footprints behind, before eventually it died. The scene was then swiftly buried by a protective layer of sediment that stopped the footprints and the dead animal washing away; the original mud hardened and eventually formed the limestone slab. 'What we see is its final death march,' Conway Morris says. He also holds up some fossilized dung from an extinct marine reptile, which reveals what this animal ate (in this case it was other baby marine reptiles).

'Palaeontology is piecing together a jigsaw,' Conway Morris says. One of the most intriguing

Conway Morris and the audience watch as a model brachiosaurus rears its head towards the ceiling of the Lecture Theatre.

puzzles is what dinosaurs were really like. 'These animals were simply titanic,' he says. To give the audience a vivid idea of just how big some of them were, a full-size model of a brachiosaurus rears its head high into the Lecture Theatre and almost touches the ceiling.

Being so enormous must have raised particular problems. For one thing, how did they get blood all the way to their heads? To find out, Conway Morris enlists the help of James from the audience who has a go at working the brachiosaurus's heart, fashioned from a hand pump. It's clearly hard work, as red liquid slowly rises through a hosepipe towards the head. But what happens to all that blood when it comes back? James finds out when he flips a switch and the blood flows down into a pair of lungs – two balloons in a glass jar, which very quickly swell and burst. 'Ghastly,' Conway Morris laments, as the audience laughs at this unfortunate event. He explains how this tells us that huge dinosaurs must have had sophisticated hearts in two parts; one was powerful and sent blood to the head, the other was less powerful and sent blood to the lungs without bursting them.

Another part of the dinosaur puzzle is how fast they moved. A volunteer on a treadmill jogs at 2.5 metres per second and his feet fall on lines spaced 1.2 m (4 ft) apart. Next, in a film clip, Linford Christie runs a 100-metre sprint at 10 metres per second with much longer strides. (In 1993, Christie became the first European to run 100 metres in under ten seconds and he still holds the British record today of 9.87 seconds; Usain Bolt holds the current world record of 9.58 seconds – 0.29 seconds faster.) For all sorts of walking and running animals – dogs, cats, ostriches, horses and others – as speed increases so does stride length. We can then work out how fast dinosaurs moved by measuring the distance between their preserved footprints. Such calculations reveal that you could easily outrun a *Tyrannosaurus rex* with a good head start. 'But watch out for the *Velociraptors*,' he warns. 'Even Linford Christie would be hard pushed to outstrip some of these very agile, small dinosaurs. They went like lightning.'

For a long time, the biggest puzzle about the dinosaurs was why they went extinct. It was only in 1980 that American physicist Luis Alvarez and his geologist son, Walter, first proposed a theory for the dinosaurs' demise that's now widely accepted. In rocks around 65 million years old, they found levels of iridium a hundred times greater than normal; this element is rare in the earth's crust but abundant in meteorites. In similar-aged rocks they found sand grains etched with characteristic fine lines. Known as shocked quartz, these crystals are also found near underground nuclear explosions. They are, Conway Morris says, 'a dead giveaway for something big and nasty'.

Luis and Walter Alvarez deduced that a meteorite, 10 km across (6 miles), slammed into the Yucatán Peninsula in Mexico around 65 million years ago (a 2013 study has pushed the date back a million years). The impact unleashed a global disaster. It sent 'mega-tsunamis' racing across the planet, flooding coastlines. It exploded into a vast fireball, setting forests on fire for thousands of kilometres, hurling soot and dust into the atmosphere, blotting out the sun for up to a year. 'The lights had been turned off,' Conway Morris says, killing off plants on land and phytoplankton in the sea. The meteorite vaporized sulphur-rich rocks, releasing plumes of sulphurous gases, which combined with water and fell back to earth as acid rain (recent studies suggest this acidified the oceans, killing off many species with calcium carbonate shells, including the ammonites). Most scientists now agree that this great catastrophe wiped out the dinosaurs and three-quarters of life on earth. 'It was millions of years before the planet came back to full health,' Conway Morris says.

This was one of five mass extinctions that have rocked the earth. 'We're now in the sixth mass extinction,' Conway Morris says. 'It's not because of meteorites, or ice ages, or other things proposed in the past,' he says. 'This mass extinction is due to us.'

Simon Conway Morris (b. 1951)

Born in Surrey, palaeontologist Simon Conway Morris has been based at the University of Cambridge since he studied for his PhD on the famous fossils of the Burgess Shale. He's a Fellow of the Royal Society. His books include *The Crucible of Creation*, about the Burgess Shale, and *Life's Solution*, in which he argues that if evolution on earth were to run again, there would ultimately still be 'sentient species with a sense of purpose'. Among their recent discoveries, Conway Morris's research group identified tiny fossils of Cambrian sea creatures with sack-like bodies, which could be the oldest ancestors of vertebrates ever found.

Long before we humans showed up, there was very little life on earth. 'Three thousand million years ago,' Conway Morris says, 'there was nothing more than bacteria.' It took a long time for complex life forms to get going. In particular, something remarkable happened 540 million years ago, when the earth suddenly filled up with an astonishing range of animals. The Cambrian Explosion, as it's known, was 'one of the great turning points in the history of life', he says. Before then, animal life had been mostly simple, unmoving creatures like sponges. Then a myriad of odd-looking animals arose with eyes and legs and swimming appendages. Palaeontologists have wrangled for decades over what triggered this evolutionary jamboree. Some say it could have been an increase in oxygen leading to the development of fast-moving, predatory animals (which need plenty of oxygen) and in turn other animals evolved into new forms to avoid getting eaten. Researchers are hunting still for answers to this ancient mystery.

Evidence for the Cambrian Explosion first came to light in 1909 when American geologist Charles Doolittle Walcott uncovered a treasure trove of fossils in the Canadian Rocky Mountains of British Columbia. A layer of rock, six feet deep, covering an area roughly the size of the RI Lecture Theatre, is known as the Burgess Shale. Walcott collected tens of thousands of what Conway Morris calls 'the most superb and fabulous fossil specimens you could ever want to study'.

He shows two Burgess Shale fossils; one looks like a lobster's tail, the other like a jellyfish, and indeed that's what Conway Morris and other palaeontologists originally thought they were. But, he confesses, those ideas were 'wildly, gloriously wrong'. Fossils found more recently in the Burgess Shale include one with all these pieces arranged in a single animal – a giant predator, up to a metre long, called *Anomalocaris*. Conway Morris shows a model reconstructing what this animal looked like. At the head end it had two 'lobster tails', which were in fact appendages used to grab prey, and the 'jellyfish' was its mouth. These strange creatures were early arthropods, ancient relatives of the myriad of arthropods alive today including the scorpion, tarantula and Madagascan hissing cockroaches that Conway Morris shows the audience.

Another fossil from the Burgess Shale is a worm-like creature, 4 cm long (1.6 in), which at first doesn't appear at all remarkable until Conway Morris points out a line running along one side. 'This is the precursor to our vertebral column,' he says. This animal is a *Pikaia*. When Walcott originally found it in 1911, he thought it was a worm, but in 1979 Conway Morris re-examined the fossils and realized it was in fact a type of chordate – the group of animals that includes all the vertebrates. This means *Pikaia* could be one of the oldest known ancestors of all the vertebrates. 'This humble little fossil,' he says, 'ultimately gave rise to such things as emus, dinosaurs, camels and, of course, us.'

The Cambrian seas were bustling with life but at that time there was nothing living on land. Around 450 million years ago, plants first moved out of the water and they were followed by early arthropods. Conway Morris shows a giant millipede's fossilized footprints left in a muddy coastal margin, which quickly hardened before they were washed away. This was one of the first invertebrate animals (without backbones) to colonize land.

Jumping forwards 100 million years he explains how palaeontologists found the answer to another great evolutionary mystery. It had long been thought that fish hauled themselves onto land on their fins, like mudskippers do today, and that this led to the evolution of terrestrial vertebrates, the tetrapods. But, as Conway Morris explains, that theory turns out to be wrong. He shows a model of *Acanthostega*, an animal that lived in water; it breathed through gills and yet had arms and legs, fingers and toes. In 1987 a team from Cambridge University led by Jennifer Clack found intact fossils of this creature in Greenland. 'I can't overemphasize how revolutionary this story is,' he says. 'Here is an animal that was all ready to go onto land.' *Acanthostega* and other early tetrapods evolved legs while living in the water, probably for clambering around aquatic vegetation. It then took only a few million years to evolve into other species with stocky arms and legs that could support themselves against gravity. These led, in time, to amphibians, reptiles, birds and mammals, including humans.

Conway Morris with a model of Acanthostega, *an extinct intermediate stage between fish and amphibians.*

The story in our bones continues as Conway Morris examines various crucial steps that took place more recently, as modern humans evolved. From roughly 6 million years ago, changing climate caused forests in Africa to be replaced by open grasslands. Around that time our hominid ancestors began walking on two feet, instead of 'knuckle walking' as our close relatives, chimpanzees, do today. He shows the fossilized footprints left by two early hominids as they strolled on two feet across the African savannah. One benefit of this upright stance may have been to keep cool. A volunteer steps into the Lecture Theatre wearing a 'Hypercolor' T-shirt that changes colour with heat (a 1990s craze in Britain) and stands under a hot, model sun. His T-shirt reveals that his body gets much hotter while he crouches on all fours compared to standing upright. Knuckle walking is fine in a forest, with a canopy casting shade, but in open grasslands there's a much greater risk of overheating. 'He'll have to drink gallons of water, which may be difficult to find,' says Conway Morris.

The bones of our early hominid ancestors also tell us what life was like back then. Conway Morris shows a piece of hominid skull punctured by two holes that fit the teeth in a leopard's skull. Whether they were hunted alive or scavenged, these early hominids were part of the ecosystem. 'You would have expected to be eaten by a large cat,' he says. But it went both ways. A picture of the cells inside a bone of *Homo erectus*, one of our early hominid cousins, shows signs of a debilitating illness caused by an overdose of vitamin A, perhaps the upshot of eating too many lions' livers. This wasn't lethal, though, and the ailing hominid survived, presumably cared for by friends and family.

Then, around 50,000 years ago, our ancestors began making increasingly sophisticated tools. As an example, Conway Morris shows a spear decorated with intricate animal carvings. This marks a revolution, he says, which continues today. Using skills and technologies that began to emerge back then, *Homo sapiens* is the only species with the ability to look into the distant past and understand where we came from and how the earth filled with life. 'So, we have unique privileges and unique responsibilities,' Conway Morris tells the audience. We must look after the planet and the remarkable things that live here, and we must continue to go out and explore it. 'History is in our bones,' he says, 'but the future is in your hands.'

66 FROM CONWAY MORRIS . . .

With so many fossils to show, from tiny, unborn dinosaurs to one of Mary Anning's famous ichthyosaurs from Lyme Regis, it took lots of preparation to make sure the Lectures ran smoothly. The television cameras couldn't be easily moved, so it was important to make sure that the objects were always in view. 'They did what we call a "stagger",' says Conway Morris, 'where you plot all the positions so the camera knows where you are.' That was done on each day of the Lectures, followed by a dress rehearsal for a few people, and then the full performance.

He commends the RI's lecture staff, Bryson Gore and Bipin Parmar, and producer Cynthia Page. 'They made it, not me,' he says. 'They were superb.' There's only one slip-up he remembers. When he was demonstrating how to weigh a dinosaur by dunking a model in water and measuring the volume it displaces, he nearly flooded the Lecture Theatre. 'The floor wasn't flat, or the trolley wasn't flat,' he says, 'and water started flowing all over the place – it was an absolute disaster. We had to stop the recording.' 99

From the archive . . .

Conway Morris borrowed a lot of fossil specimens from various museums, including the Natural History Museum in London. This invoice details resin casts of fossils he used in his Lectures, including a Neanderthal skull and a harpoon point.

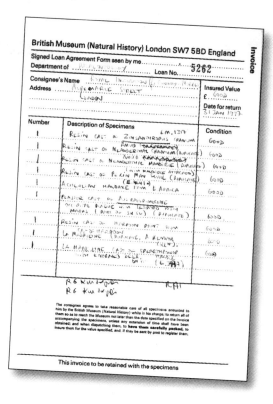

This invoice to be retained with the specimens

CHAPTER 10

To the End of the Earth: Surviving Antarctic Extremes

Lloyd Peck

2004

✧

For the first time in the history of the Christmas Lectures, the audience is transported away from London and down to the vast frozen continent at the bottom of the world. Featuring footage filmed specially in Antarctica, Peck unlocks the secrets of how life survives in the coldest place on earth. It's a long way away but, as Peck reveals, Antarctica's fragile, ice-bound realm is critical to each and every one of us. And in another first for the Christmas Lectures we hear in detail about the growing problem of climate change.

✧

The floor of the Lecture Theatre is filled with a giant, snow-covered model of the Antarctic continent. There's just enough room left for a chest freezer. The lid flips open and out climbs Peck. 'The temperature in this deep freeze is minus twenty degrees,' he says. 'Who would be foolish enough to live in an environment like that?'

On a large screen, the audience sees a formidable white landscape of ice fields, snow-covered peaks, icebergs and dark ocean waters. Peck, now kitted out in scuba gear, pops his head up from the sea. 'I'm an Antarctic marine biologist,' he says, 'and this is my office.' The reason biologists like Peck go to the Antarctic and endure freezing conditions is to study the extraordinary wildlife that lives there. 'This frozen frontier is home to some of the toughest, most well-adapted creatures on the planet,' he says.

Peck has brought a selection of live marine animals back with him from Antarctica to show the RI audience, comparing them to more familiar creatures. A common garden woodlouse wanders over Peck's hand. It's a type of animal known as an isopod, a relative of crabs and lobsters. 'There are isopods in Antarctica,' he says, 'but there's one big difference.' He picks up a pair of enormous isopods, each one as big as his hand, with spiky, wriggling legs. The audience gasps excitedly. 'Keep them away from me!' squeals one boy.

'Now I'm going to show you an even cooler example,' Peck says. The small blob on his fingertip is a full-grown example of one of Europe's biggest sea spiders (relatives of the horseshoe crab that Julian Huxley brought into the RI Lecture Theatre in 1937; see page 53). There's more giggling and chattering as Peck brings out a red Antarctic sea spider, draped over his hand. 'The biggest ones are the size of a dinner plate,' he says.

Peck shows his audience a live giant sea spider from Antarctica.

To meet more of Antarctica's giants, Peck takes his audience on a dive into the world's coldest ocean. On screen, Peck jumps through a hole in the ice. 'I really enjoy diving here,' he says, 'despite the cold.' It's around -2°C down there. He swims past sea urchins and enormous, forty-armed starfish, all cold-blooded creatures (known as ectotherms) whose body temperatures match their surroundings. They don't freeze because they're very salty, which lowers their freezing points (it's also why the sea doesn't freeze here). Antarctic fish make their own anti-freeze to stop themselves from turning into blocks of ice. These molecules, a type of glycoprotein, circulate in the blood stream and prevent ice crystals from growing bigger.

Back in the Lecture Theatre, Peck demonstrates an important property of cold water. He has two beakers of water with a coiled heating element in them. One is as at -2°C and the element is covered in silvery bubbles of oxygen coming out of solution. The other beaker, at a tropical 30°C, has no bubbles. The colder water contains more oxygen because the gas takes up less space, and more can fit between the water molecules than when it's warm. Peck demonstrates this by plunging a balloon into liquid nitrogen; as the air cools inside, the balloon deflates. It's thanks to the cold, oxygen-rich water that Antarctic animals can grow so big. Things like sea spiders and starfish rely on oxygen diffusing around their bodies; for big animals this only works if there's plenty of oxygen.

Warm-blooded animals (endotherms) living in Antarctic seas generally keep warm by being extremely fat. On the screen, the audience sees Peck strolling across a beach in the Falkland Islands, north of Antarctica. He's surrounded by enormous elephant seals. 'Bulls like this one weigh in at over three and a half tonnes,' he says, carefully keeping

Peck filming with elephant seals on location in the Falkland Islands.

his distance. 'They're built like this not just for insulation but because they need to fight.' Two males rear up and bellow at each other. The biggest seals defend the biggest harems of females and sire lots of pups, which are also very chubby. They triple in weight in twenty days, before migrating south to the cold waters of Antarctica.

On land, conditions in Antarctica are even more extreme than underwater. Around 99 per cent of the continent is covered in snow and ice. Male emperor penguins tough it out for ten weeks, at -50°C, rearing their chicks on the ice. They huddle together, taking turns to face the full force of the wind; 40 per cent of their body weight is fat.

No large animals live permanently on land in Antarctica. There are no polar bears or foxes, partly because the southern continent is so isolated. Unless they can swim or fly long distances, there's no way for animals to escape when conditions get really inhospitable. In contrast, in the Arctic, the Asian and North America continents are near enough that animals can cross the sea ice and walk to warmer climes when they need to. Some of Antarctica's biggest animals living permanently on land are tiny insects called springtails.

But it hasn't always been this way. Millions of years ago, Antarctica was much warmer and filled with life, including lush forests. Peck shows the audience a dinosaur skull. It was a herbivore, about three feet high, that roamed Antarctica 75 million years ago. 'Like other Antarctic dinosaurs, it had really big eyes,' Peck says. That's because the continent was over the South Pole, in the same place as today, and there were long dark winters.

Why did Antarctica go from being warm to cold? Peck explains that it used to be attached to South America and a warm current flowed down from the tropics, preventing Antarctica from freezing. Then, 35 million years ago, South America drifted north. 'Antarctica became isolated,' Peck explains. 'Without the influence of the warming currents, snow and ice began to gather on top of the cold land.'

Today, 30 million cubic kilometres of ice sit on top of Antarctica, weighing 30 quadrillion tonnes. Its weight distorts the planet, pushing the continent two-thirds of a mile into the earth. The ice moves in vast networks of glaciers, over four kilometres deep. On the screen, the audience sees Peck abseiling down into the icy blue heart of an Antarctic glacier. Inside a glacier, the pressure of so much ice piled up causes a thin layer at the base to melt, and the whole thing slides along at up to ten metres a day. Glaciers slide outwards from the centre of the Antarctic continent, because the land underneath is shaped like an upturned bowl. Reaching the sea, they either break up into icebergs or spread onto the sea as floating ice shelves that are still connected to land.

'But in parts of Antarctica,' Peck says, 'this world of ice is changing, and it's changing quickly.' Temperatures in the Antarctic peninsula are rising faster than anywhere on earth. 'They've gone up

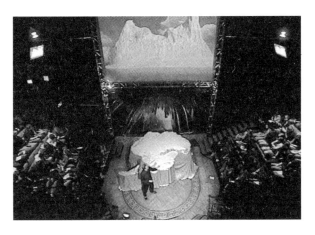

A giant model of Antarctica fills up much of the floor of the Lecture Theatre as Peck introduces the audience to this white, frozen continent.

three degrees in fifty years,' he says. 'This is having a tremendous effect on the ice.' Peck shows images of the Larsen B ice shelf, which held 720 billion tonnes of ice. In 1992 it broke away from Antarctica and floated away. 'The collapse of the Larsen B ice shelf is the clearest indication yet that the world is warming.'

Lloyd Peck (b. 1957)

Born in Walsall, in the West Midlands, Peck studied natural sciences at Jesus College, Cambridge, then worked for a year in the steel foundry back home before returning to academia, and doing his PhD at Portsmouth University on growth and reproduction in marine snails. In 1984 he joined the British Antarctic Survey in Cambridge, where he's professor and science leader of the Biodiversity, Evolution and Adaptation team. Peck is also a visiting professor at the universities of Sunderland and Portsmouth, an honorary lecturer at the University of Cambridge and a Scientific Fellow at the Zoological Society of London. Following his 2004 Christmas Lectures, he gave them again for televised series in Japan, Korea and Brazil.

Evidence for climate change comes from the Antarctic ice itself. Ice cores drilled from glaciers contain layers of snow that fell and formed ice over the last 900,000 years. 'You can go to any year in this frozen calendar and find out what the atmosphere was like,' Peck says, showing his audience a slice from a deep ice core. It's full of tiny bubbles that were trapped when the ice formed thousands of years ago. Scientists measure carbon dioxide levels in those trapped bubbles. They can also work out what the temperature was by measuring two forms of oxygen (known as isotopes) in the ice: oxygen 16 and oxygen 18, a heavier and rarer version. In warm conditions, when the water has more energy, more of the heavier oxygen 18 evaporates and falls as snow. By measuring ratios of oxygen 16 to 18, scientists calculate past temperatures. Antarctic ice cores show that in the past, as carbon dioxide levels went up, so did air temperature. 'Carbon dioxide in the atmosphere acts like a one-way insulator,' Peck says. 'It allows in heat from the sun but doesn't let it all back out again. The really worrying thing is that carbon dioxide levels are higher and rising faster than we have ever seen.'

Scientists prepare an ice core in Antarctica.

Peck shows a film of him visiting the Devonshire home of James Lovelock, the independent scientist who proposed the Gaia hypothesis, the theory that the earth is a self-regulating system. 'Sometime during this century we will pass a threshold level of carbon dioxide in the atmosphere which is somewhere in the region of 500 parts per million,' Lovelock says. 'When we pass the threshold the system is then committed to change and nothing we do after that will make the slightest difference; it will move of its own accord, into a state much warmer, much hotter than we've ever experienced.' In 2017 the earth's atmosphere reached 410 parts per million (ppm) of carbon dioxide for the first time in millions of years (in 1958 it was 280 ppm), and none of us are likely to see it fall back below that level in our lifetime. 'This was one of the most worrying conversations I've ever had,' says Peck, back in the Lecture Theatre. 'What is clear is how short a time we have to act.'

To find out how much of a role members of his audience are playing in this, Peck hands out the following questionnaire for everyone to fill in, gauging their individual impact on the planet (the numbers give a relative score, combining an estimate both of the resources consumed and the carbon dioxide released in daily life):

What do you eat?
Fresh food (5), processed food (10), ready meals (15)

Where do you live?
Flat (5), terraced house (15), detached house (35)

Where did you go on holiday last year?
UK (10), Europe (20), rest of the world (150)

How do you travel each day?
Walk or cycle (3), public transport (25), car (50)

How often do you wash?
Bath every day (40), shower every day (20), shower every other day (2)

The average score for people in Europe is 50–100. In the US, it's over 200. If everyone alive today consumed at American levels there simply would not be enough resources to go round; to support a global population at those levels of consumption we would immediately need roughly five planet's worth of resources, which obviously aren't available.

Rising temperature is already having a major impact on Antarctic life. On the screen, Peck climbs up a rocky outcrop in Antarctica and points out a green tuft of grass. 'Fifty years ago there wasn't any grass growing here,' he says. Now plants are spreading across the Antarctic peninsula, from seeds blown in from South America. Seeds only germinate on bare rock, so as the peninsula warms and ice melts, more rock is becoming exposed for plants to colonize. 'This isn't just going to stop at grass,' Peck says. 'Small insects and beetles are probably going to come next, and they will have a devastating effect on that fragile land ecosystem.'

At the British Antarctic Survey, Peck's research team studies how sea creatures respond to rising temperatures, including clams that live burrowed in the Antarctic seabed. In a time-lapse video, we see these huge clams in an aquarium tank. At zero degrees, it takes them twelve hours to dig down and rebury themselves in the mud. At five degrees they don't move at all. Around 5,000 cold-blooded species live in Antarctic seas. If even a small fraction of them are as sensitive as the clams, Peck warns, it could have a catastrophic effect on marine ecosystems.

As well as impacting wildlife, rising temperatures in Antarctica could profoundly impact the rest of the planet. First, Antarctica plays a key part in the world's climate because of a system of ocean currents called the global conveyor belt. In the Lecture Theatre, Peck shows with a tank of water how this works. He adds a lump of blue ice to one end; the blue water sinks to the bottom. This represents the sea ice in Antarctica, which locks up pure water and leaves the salt behind, making the sea saltier and denser, causing it to sink. Peck then adds warm, red water that's less dense and floats on the surface, representing warm water at the equator. This sets up a circulation, like the global conveyor belt, which moves water across the oceans and influences climate. 'Warm currents bring warm, wet weather,'

Peck demonstrates how the global ocean conveyor belt works with a tank of water, ice and coloured water.

he says, 'cold currents cause cold, dry conditions.' The problem, Peck explains, is that sea ice in Antarctica has decreased by 20 per cent in the last fifty years. It's not clear yet what impact this will have on the global conveyor belt, but there's no doubt it's vital for the climate.

Warming in Antarctica could also cause sea levels to rise. 'Eighty per cent of the world's freshwater is in Antarctica's ice,' says Peck. And with rising temperatures, the ice shelves could give way, just like Larsen B did (in fact, in July 2017 a trillion-tonne iceberg broke away from the Larsen C ice shelf). Ice shelves act as giant plugs that hold glaciers back. 'Remove them and these massive rivers of ice would drain the ice sheets straight into the oceans,' says Peck.

He goes on to paint a stark picture of what will happen if Antarctica's ice does melt. If just the West Antarctic ice sheet collapses, Peck says, sea levels will rise by six metres worldwide. 'All low-lying areas including major cities like New York, Tokyo and Mumbai would slowly disappear.' Half of Bangladesh would be under water.

'It's still not clear just how much ice will melt or how much the sea level will rise,' Peck says, 'but it's estimated just an 80 centimetre rise in sea level would cause twenty to thirty million refugees in India and Pakistan alone. So, rising sea level is a real threat to our civilization.'

As he brings his Lectures to a close, Peck leaves his young audience with much to think about. 'Even a small change in this fragile continent could have dramatic consequences for the human race and for the whole planet.' And it's clear we don't have long to act. 'We need to think about the way we live and change what we do. We need an army of people who can pass the message on.'

 FROM LLOYD PECK . . .

'The Antarctic filming nearly didn't happen,' says Peck in the tearoom at the British Antarctic Survey (BAS), thirteen years after he gave his Christmas Lectures. The filming had to fit into a narrow window between the end of the Antarctic winter (when it's impossible to travel in or out) and the rehearsing schedule for the Christmas Lectures. The film crew found out the hard way just how extreme conditions can be in Antarctica. They flew to the Falkland Islands and got stuck for ten days due to bad weather, with no flights going further south. There they did all the filming they could, including the beach scene with the elephant seals. Then, finally, the weather cleared and they had to squeeze the filming into a few days. 'We literally went in on the last day possible for us to get into the Antarctic,' Peck explains.

Peck has now been to Antarctica seventeen times (and the Arctic just twice). I ask him whether he's seen the influence of climate change in his time visiting and studying the icy, southern continent. 'You can see big ecological change,' he says, such as shifting penguin colonies and brittlestars (relatives of starfish) that no longer breed during warm summers. He's also witnessed the shrinking glaciers. 'The big glacier near the station at Rothera,' he says (BAS's research station), 'I've been going since 96 and that's gone back three kilometres. So, I can now go in an inflatable boat and I can sit in open water where there was a two-hundred-and fifty-foot ice cliff.'

CHAPTER 11

THE 300-MILLION-YEAR WAR

Sue Hartley

2009

✧

For 300 million years, plants and animals have lived together on land and an epic battle has played out between them. Lots of animals eat plants and in response plants have evolved ingenious weapons to protect themselves, communicate with each other, tailor their weapons to particular attackers and even summon help from other creatures. In her dynamic and often humorous Lectures, Hartley reveals how ecosystems are made of intricate and fragile webs of connections between animals and plants, ones on which we depend in many ways.

✧

A sparkling Christmas tree sits in the middle of the Lecture Theatre. 'It's very well prepared to fight the 300-million-year war,' says Hartley. She lets a few members of audience feel the sharp leaves and sniff the piney scent. 'It's weapons like these smells that make plants the most successful organisms on the planet,' she says. Compared to animals, there's a thousand times more plant material on earth, enough to fill the O2 Arena 1.5 million times. Plants are also the longest-lived and biggest living things. Hartley shows photographs of a 4,500-year-old

A giant caterpillar has trouble walking across a giant, slippery holly leaf as Hartley introduces the 300-million-year war between plants and animals.

bristlecone pine tree and a giant redwood tree that's the size of forty whales.

Some plants eat animals. Hartley shows the audience a pitcher plant that attracts insects with nectar and drowns them in a pool of digestive enzymes. Mostly, of course, it's the other way round: animals eat plants. In response, plants have evolved ingenious defensive weapons to deter these herbivores. There are physical defences like grasses covered in sharp silica structures (called phytoliths), which look like microscopic shards of glass. Plants also have chemical weapons. Hartley holds up a red Anaheim pepper which, she says, has 2,500 Scoville Heat Units, or SHUs, a measure of its fiery heat. A brave volunteer from the audience, Alistair, comes down. He doesn't eat the chilli but sips water infused with just a few grams and the audience giggle while he hops about, fanning his mouth.

Next, Hartley shows a ghost chilli plant, one of world's hottest chillies with a million SHUs. The audience murmurs, no doubt expecting she'll ask someone to try it. In fact she welcomes a special guest, Adam Stacey, who's dressed in sunglasses and Hawaiian shirt. Adam chews and swallows a small ghost chilli and very soon he's unable to speak, tears streaming down his face.

While Adam suffers, Hartley explains how chillies are filled with chemicals known as capsinoids, which bind to receptors in the mouth. She then introduces another special guest – Barry, a yellow budgerigar. 'Barry is a champion chilli eater,' she says. While Adam made the mistake of chewing and crushing the potent seeds, Barry swallows them without releasing the capsinoids. Birds help plants by dispersing seeds unharmed in their droppings. 'So, the chilli likes Barry, and hates Adam,' she says. But Adam clearly hates the chilli more and swears he'll never eat one again as he hobbles off to cheers from the audience. 'I think the defence has worked,' says Hartley.

Adam suffering after chewing on one of the world's hottest chillies.

As well as red-hot chilli peppers, plants collectively make an estimated 100,000 different defensive chemicals to discourage herbivores from eating them (except the animals that disperse their seeds). These deterrents include the deadly chemicals cyanide, strychnine and ricin. Yet poisons aren't the only problems that herbivores face. Their food can also be low in important nutrients. Hartley welcomes in Yogi, a Saint Bernard dog (a meat eater) and Jerry, a Shetland pony (a herbivore), to demonstrate the different diets of these similar-sized animals. Yogi eats smaller amounts of protein-rich dog biscuits, while Jerry chews through huge piles of protein-poor hay.

Next, in comes a large model of a lethal plant-eating machine. This is Bessy the cow. 'Cows have some of the most elaborate adaptations to eating plants,' Hartley says. With the help of Lochlan from the audience, she shows how Bessy rips up grass with a dextrous tongue and shovels it down her throat. Next, six huge containers of water are wheeled in to show how much saliva cows produce every day – 110 litres – to help wash down a daily dose of 70 pounds (over 30 kg) of grass. Lochlan then opens up Bessy's stomach to reveal four chambers. The first, the rumen, contains the cows' secret weapons: billions of bacteria that digest cellulose, the tough part of plant cell walls. In the second chamber, bacteria keep doing their work but the pieces of grass are too big to break down completely. Bessy has to chew the grass a second time. 'She's got to bring it back up!' says Hartley grimacing, explaining the importance of chewing the cud. The third chamber absorbs water and the fourth is more like a regular stomach (including that in humans) and contains hydrochloric acid and digestives enzymes to break the food down.

Back inside Bessy we see she has another weapon in the war against plants; she has very long, impressive intestines represented by a thick rope, which Hartley passes into the audience, looping it around the room. Compared to humans, with a mere six metres of intestines, a cow's stretch for fifty metres, all the better for absorbing nutrients. At the end of the lengthy intestines, one lucky audience member, Adam, gets to come down and lift up Bessy's tail to reveal the inevitable – all the indigestible grass remains come out as a large cowpat.

In the war between plants and animals, communication plays a vital role. Even though they can't speak or see, plants are communicating all the time.

Hartley and her helpers bring on two donkeys to demonstrate how herbivores adapt to eating plant material.

To demonstrate how, Hartley transforms the audience into a forest of trees. Under each seat is a pot of bubble mixture. 'When you see a bubble, or one lands on you, then you start blowing,' she instructs. People sitting in the very back row set off a cascade of signals that sweeps forwards and soon the Lecture Theatre is full of bubbles. 'That's how plants can communicate with each other, and warn each other,' she says.

To explain this in more detail, Hartley describes experiments she did as a student when she discovered that plants know when particular herbivores are attacking. The key is in caterpillar spit. First she snipped plant leaves with scissors – imitating caterpillar jaws – and found this wasn't good at triggering a plant's chemical defences. Then she extracted caterpillar saliva. 'I found that if you hold caterpillars carefully they act like giant tubes of toothpaste,' she says. If she smeared saliva on the snipped leaves she got the same response as if a real caterpillar was chewing the plant: it began making poisons. Caterpillar saliva also makes plants emit their own airborne chemical signals (as represented by the bubbles), warning other plants to prepare for attack and start producing anti-herbivore poisons.

As well as talking to each other, plants also talk to animals. Wasps smell the plants' warning signals and fly in to investigate. To demonstrate what happens next, Hartley introduces Kenny, a model caterpillar, and invites a volunteer from the audience, Jack, to pull off Kenny's head. 'Oh,' Hartley cries, 'you've killed Kenny!' Jack shoves his hand into Kenny's decapitated neck and pulls out a handful of sticky goo and giant, model grubs. Inside the caterpillar, a wasp laid hundreds of eggs by piercing through its skin with a sharp egg-laying needle (called an ovipositor). The eggs then hatched and started eating

Hartley and Kenny the caterpillar before a member of the audience looks inside and finds it's full of parasites and goo.

Kenny from the inside out. 'Plants can signal to the wasps,' says Hartley, 'and then its curtains for Kenny.' Parasitic wasps usually attack only one specific kind of caterpillar, so plants customize their messages. 'They can send signals to exactly the right sort of wasp,' she says.

So far, the 300-million-year war has been well balanced between plants and animals. However, Hartley says, that careful balance could be upset by climate change. Shifts in temperature and rainfall can alter the nature of the battle. 'There are examples of this happening now,' she says. 'It's rare but we do see sudden increases in insect numbers that seem to overwhelm plant defences.' She shows pictures of plagues of locusts devouring crops.

'Are we going to see more insect outbreaks in the future?' Hartley asks. One group of insects to watch are aphids, which have sharp mouth-parts that pierce plants and suck out their juices, like tiny vampires. She brings out an aphid-infested plant, covered in black dots. These aphids, she warns, are among the most dangerous pests, causing £100 million of damage to cereal crops every year. She shows why they're such a problem with the help of a giant model aphid, called Angie. If all Angie's offspring survived, Hartley explains, there would be a layer of aphids covering the earth 150 km deep, reaching half the way to the International Space Station. This comes down to the way aphids breed without having sex.

Another volunteer, Conor, comes down to help deliver Angie's babies. Inside her he finds a cloned version of Angie, called Alice. But that's not all. Inside Alice there's another cloned aphid, called Alison. When Angie was pregnant, Alice inside her was already pregnant with Alison, a system known as telescopic generations. 'This amazing way of breeding,' Hartley says, 'allows aphids to reproduce

Hartley with a model of Angie the aphid.

very quickly.' For now, predators, including ladybirds, are keeping aphids in check. But as the world warms, aphids will reproduce even faster until eventually predators may get swamped and, as Hartley puts it, we could drown in aphids.

As temperatures rise, some plants won't be able to summon help from insect predators. Horse chestnut trees (also known as conker trees) are one such plant at risk from climate change. Across Britain, by late summer many horse chestnut tree leaves look blotchy and brown. Hartley introduces the culprits – fluffy silver-and-brown-striped moths called horse chestnut leaf miners. These caterpillars have huge appetites for horse chestnut leaves, munching them from the inside out. 'There can be up to 700 miners on a single leaf,' she says. In their native range in Italy these caterpillars are kept in check by wasps (a parasite that lays eggs inside the moth caterpillars), but they haven't made it to the UK. 'The miner has spread north without its enemies,' she says.

Climate change may be tipping the balance of the 300-million-year war in favour of insects, but plants have one final secret weapon. 'Seeds,' Hartley says, 'are like little time capsules that preserve plants for the future.' She welcomes in seed expert, Wolfgang Stuppy, from the Royal Botanic Gardens at Kew. Stuppy shows microscope images of intricate seeds. Some look like brains and honeycombs, some are covered in spikes and hooks. 'Under a microscope they look very scary,' he says. Plants use animals to disperse their seeds, either carried in their stomachs (like the budgie eating chilli seeds) or hooked on their fur and feathers. Some seeds blow around in the wind. 'These strategies allow plants to spread their seeds far and wide,' says Hartley.

Seeds also contain amazing genetic diversity that could provide solutions to problems in the human world, including crops that will survive as the climate changes. Hartley shows a film of her recent visit to Kew's Millennium Seed Bank, which she describes as 'Noah's Ark for plants'. She visits a series of cold, underground, concrete bunkers with rows of huge glass jars storing 2 billion seeds collected from 10 per cent of the world's plant species. (Kew's next aim is to conserve 25 per cent of the world's plant diversity by 2020.) Kept at -20° degrees, many seeds survive for thousands of years.

Back in the Lecture Theatre, Hartley introduces two final guests that come from somewhere plants and animals have lived together for millions of years in isolation from the rest of the world. 'This means they've ended up very weird and wonderful,' she says. First, there's what looks like a spindly cactus but that in fact comes from a completely different family that's only found on one island. This is the ocotillo of Madagascar. It has tiny green leaves growing between long spines. 'Eating these must be really hard,' Hartley says. One group of Madagascar's animals has evolved to do just that. In comes Curtis, a ring-tailed lemur. Enchanted oohs and aahs ripple through the audience.

Sue Hartley (b. 1962)

Hartley studied biochemistry at Oxford University and received her PhD from the University of York, before moving to the University of Sussex where she was Professor of Ecology when she gave her Christmas Lectures. After the televised shows, she gave her Lectures again in Japan and at Manchester's Big Bang Science Fair in 2010. She's now Director of the Environmental Sustainability Institute at the University of York, where she continues her studies of the interactions between plants and animals, including finding ways to improve crop resilience to climate change and using natural plant chemicals to defend crops against pests.

'Curtis has amazing hands,' says Hartley, who's clearly as besotted with the lemur as Sir David Attenborough was in his 1973 Lecture (see page 100). Kat and Suzy are the two lucky audience members who get to feed Curtis. And just like in Attenborough's Lecture, we see the lemur is very fond of grapes. 'Curtis is very good at grabbing hold of things,' says Hartley, as the lemur reaches out and opens Kat's hand for more food. In southern Madagascar, Hartley explains, lives another lemur (known as the sifaka) that evolved hands exactly the right size to fit between the spines of the ocotillo tree, to pick off and eat the leaves. 'So in Madagascar, the 300-million-year war has produced some amazing plants,' she says, 'and some amazing animals.'

Hartley ends with a powerful and heartfelt message for her young audience. 'We're going to need to be as ingenious as plants if we're going to meet the challenges of the future,' she tells them. There's no shortage of problems from climate change to food shortages. Together, plants and science will be vital in solving them. 'There's never been a time when science has been more important for our very survival,' she says. 'You guys are the scientists of the future. Your planet and everything on it is depending on you.'

From the press . . .

In an interview with the *Daily Telegraph* (18 December 2009), RI science demonstration technician Andy Marmery revealed some of the seasonal problems he faced in staging Hartley's Lectures: 'We haven't done plants in the lectures since the 1830s, and now I know why: they don't like Christmas. Most of them are dead or leafless, which is making my props a bit of a challenge.'

Science writer Ed Yong was in the audience for Hartley's Lectures. 'There is something utterly uplifting about seeing a group of kids running, nay, fighting their way up a set of stairs to get to a science lecture', he writes on 6 December 2009 in his blog *Not Exactly Rocket Science*. 'So well judged is Hartley's talk that they lap up every word with rapt attention. When she asks for volunteers, the kids go *mental*.'

 ## FROM SUE HARTLEY . . .

'It was a huge amount of fun,' says Hartley, looking back on her Christmas Lectures. 'It was really all about the kids and their excitement and interest and learning experience.' Her Lectures were packed with models of animals and plants. Did she get to take any of them home with her? 'Kenny the caterpillar, with the colourful tail and the grubs inside . . .' she says, 'he's in my office.' There were real animals, too,

Hartley poses with the two donkeys who appeared in her Lectures.

which weren't always well behaved. One herbivore the audience met was Calypso, a two-toed sloth. Beforehand, she was quite active, but as soon as the cameras started rolling she crawled to the side of the stage and sat very still. 'At one point the director was telling the floor manager, "All I can see is its flipping arse,"' Hartley recalls.

She was the fourth woman to give the Christmas Lectures and only the second botanist. The first botanist to give the Lectures, John Lindley, who gave the 1833–4 Christmas Lectures, played a vital role in saving London's Kew Gardens when the government threatened to close them down in the early nineteenth century. Now Hartley is on the board of trustees at Kew and she remains a passionate advocate for the often-neglected plants and the urgent need to protect them.

At the end of her last Lecture, Hartley seemed to get a little tearful as she delivered her poignant message about the state of the planet. 'It did just come home to me that we were talking about something that might well be lost . . . all this amazing biodiversity,' she says. 'So it did feel like a very emotional ending. It was an amazing experience, I think that was probably part of it as well . . . it had been such a powerful thing to do and we ended on that very powerful message.'

EPILOGUE

Today, we understand more than ever before about the inner workings of the natural world. Over the course of a century of Christmas Lectures, we've seen how biologists have constantly revealed more details about species and ecosystems, from the ocean depths to mountaintops and everywhere in between. We also know more than ever about how human activities are upsetting the delicate balance of the web of life. Some problems have been solved, while new ones have emerged.

A few key messages thread all the way through these Lectures. One is how important the natural world is for human lives, although we often don't realize this until things go wrong and the balance is lost, when pests and diseases take over. Another important message is that a multitude of immense natural wonders are out there for any of us to find – whether it's the fossilized remains of an ancient spiralling shell, a ravenous, carnivorous beetle in a garden pond or a bee carrying around a flower's pollen with the purpose (perhaps) of making more flowers.

More than half of the human population now lives in cities and many children are growing up with little contact with nature. There's never been a more important time to find ways for people to reconnect with the natural world and to know and care about what's out there. The RI Christmas Lectures play a vital part in bringing nature into vivid view for so many people and nurturing a sense of curiosity, encouraging everyone to think about the living world in new ways, and simply to go out and explore it.

Author's Note

My warm thanks go to everyone who's helped with this book, especially to Gail Cardew, Charlotte New and Liina Hultgren at the RI for all their help in contacting living Lecturers, guiding me through the RI's archives and providing film footage of the Lectures. Many thanks to Jane Acred at the Department of Zoology in Cambridge for helping me track down James Gray's archives. Huge thanks also to Jo Stansall at Michael O'Mara Books for nurturing the book all along the way. Special thanks to Desmond Morris, Robert Attenborough, Simon Conway Morris, Lloyd Peck and Sue Hartley for kindly sharing your recollections of the Lectures with me; it was a pleasure to be in touch with you. To all the Lecturers who have given me so much to think about, not only in writing this book but also over years of being a fan of the Christmas Lectures – thank you all. And Ivan, thank you for your continued love and support throughout all my wordy endeavours.

Here are brief details of the source materials used for each chapter:

Chapter 1: The Childhood of Animals

Peter Chalmers Mitchell turned his Christmas Lectures into a book of the same name, published in 1912 by Cambridge University Press. Direct quotes are taken from that and newspaper accounts of the Lectures.

Chapter 2: The Haunts of Life

Direct quotes come from John Arthur Thomson's 1921 book (of the same name) published by A. Melrose Ltd, and also newspaper accounts of the Lectures.

Chapter 3: Concerning the Habits of Insects

Newspaper accounts of the Lectures and Francis Balfour-Browne's 1925 book (of the same name), published by Cambridge University Press, provided the direct quotes.

Chapter 4: Rare Animals and the Disappearance of Wild Life

Details of Julian Huxley's Lectures were pieced together from the official RI programme for the Lectures and from various newspaper reports.

Chapter 5: How Animals Move

Direct quotes come from James Gray's book based on his Lectures (Cambridge University Press, 1953) as well as newspaper reports on the Lectures, which quote him directly.

Chapter 6: Animal Behaviour

No film footage is available of these Lectures. Desmond Morris kindly provided his recollections to me via email. Other details were taken from the RI's programme for the Lectures.

Chapter 7: The Languages of Animals

Video footage exists of David Attenborough's Lectures (except for Lecture 4; a written transcript of this is available in the RI archives). All the direct quotes were taken from the films.

Chapters 8–11: Growing Up in the Universe; The History in Our Bones; To the End of the Earth: Surviving Antarctic Extremes; and The 300-million-year War

Direct quotes from Richard Dawkins, Simon Conway Morris, Lloyd Peck and Sue Hartley were all taken from video footage for each of these Lecture series.

PICTURE CREDITS

Page 1: Photo of the 2016 Lectures, given by Professor Saiful Islam; Paul Wilkinson Photography

Page 6: Lecture programme (front cover); from the collection of the Royal Institution (RI MS AD 06/A/03/A/1911)

Page 9: Various stages of prawn larvae; illustration from original edition of *The Childhood of Animals*, Peter Chalmers Mitchell, Frederick A. Stokes Company 1912

Page 11: Metamorphoses of an axolotl; illustration from the original edition of *The Childhood of Animals* (see above)

Page 17: Photo of Mitchell; © Hulton-Deutsch Collection / Corbis via Getty Images

Page 19: Family of lions plate; illustration from the original edition of *The Childhood of Animals* (see above)

Page 22: Lecture programme (front cover); from the collection of the Royal Institution (RI MS AD 06/A/03/A/1920)

Page 23: Illustration of Thomson giving his Christmas Lectures; *London Illustrated News*, 15 January 1921

Page 28: Sea lilies (crinoids); illustration from the original edition of *The Haunts of Life*, J. Arthur Thomson, Harcourt, Brace & Company 1922

Page 32: A velvet worm; illustration from the original edition of *The Haunts of Life* (see previous page)

Page 35: Gossamer spiders; illustration from the original edition of *The Haunts of Life* (see previous page)

Page 36: The school of the shore; illustration to accompany Thomson's Lectures, *London Illustrated News* 1921

Page 38: Lecture programme (front cover); from the collection of the Royal Institution (RI MS AD 06/A/03/A/1924)

Page 41: Bee wall photo and bee varieties; from original edition of *Concerning the Habits of Insects*, Francis Balfour-Browne, Cambridge University Press 1925

Page 44: Photo of Balfour-Browne after a Lecture; *London Illustrated News*, January 1925

Page 46: Photo of Balfour-Browne; family portrait courtesy of the Balfour-Browne Club

Page 48: Diagram of the head of a water beetle larva and drawings of the feeding habits of water beetle larvae; from original edition of *Concerning the Habits of Insects* (see above)

Page 50: Photo of Balfour-Browne during his Lectures; *Sphere* newspaper, 3 January 1925

Page 55: Photo of Huxley and Max the lion; Royal Institution / Science Photo Library

Page 57: Photo of Huxley and parrot; Popperphoto / Getty Images

Page 58: Lecture programme (front cover); from the collection of the Royal Institution (RI MS AD 06/A/03/A/1937)

Page 62: Cave drawing of wild horse from Lecture programme; from the collection of the Royal Institution (RI MS AD 06/A/03/A/1937)

Page 69: Photo of Gray and cheetah; Royal Institution / Keystone

Page 71: Lecture programme (front cover); from the collection of the Royal Institution (RI MS AD 06/A/03/A/1951)

Page 72: Diagram of leech bridge; illustration from original edition of *How Animals Move*, James Gray, Cambridge University Press 1953

Page 73: Photo of Gray and mechanical duck; Royal Institution / Keystone

Page 76: Gray's handwritten notes on Lecture slides and Lecture 1; reproduced by kind permission of the Syndics of Cambridge University Library (Add. 8125 D19/Add. 8125 D22)

Page 79: Diagram of a galloping horse; illustration from original edition of *How Animals Move* (see above)

Page 81: Diagram of birds rising upwards air on currents; illustration from original edition of *How Animals Move* (see above)

Page 84: Letter from London Zoo to Gray; reproduced by kind permission of the Syndics of Cambridge University Library (Add. 8125 D18)

Page 87: Lecture programme (front cover); from the collection of the Royal Institution (RI MS AD 06/A/03/A/1964)

Page 88: Photo of Morris with chimp; Keystone / Hulton Archive / Getty Images

Page 93: Photo of Morris and chimp standing; Sport & General / Alpha Library

Page 94: Photo of grape shy demonstration; Reg Speller / Fox Photos / Hulton Archive / Getty Images

Page 98: Letter from Morris to Sir Lawrence Bragg; Royal Institution / Desmond Morris (RI MS AD 06/A/03/C/1964/folder 2)

Page 98: Letter from Bragg to Morris; from the collection of the Royal Institution (RI MS AD 06/A/03/C/1964)

Page 101: Lecture programme (front cover); from the collection of the Royal Institution (RI MS AD 06/A/03/A/1973)

Page 104: Photo of Attenborough and chameleon; Royal Institution / BBC

Page 106: Letter from Porter to British Museum; from the collection of the Royal Institution (NCUAS C 1085)

Page 108: Photo of Attenborough and lemur; Royal Institution / BBC

Page 111: Photo of Attenborough talking to audience; Royal Institution / BBC

Page 116 and 117: Handwritten letter from Attenborough to Porter; Royal Institution / Sir David Attenborough (NCUAS C 1085/5 Jan 1973)

Page 118: Letter from Porter to Attenborough; from the collection of the Royal Institution (NCUAS C 1085/7 Jan 1974)

Page 119: Letter from Attenborough to Porter; from the collection of the Royal Institution (NCUAS C 1085/12 Jan 1974)

Page 121: Lecture programme (front cover); from the collection of the Royal Institution (RI MS AD 06/A/03/A/1991)

Page 124: Still from Lecture 2; BBC Motion Gallery / Getty Images

Page 129: Still from Lecture 3; BBC Motion Gallery / Getty Images

Page 133: Still from Lecture 3; BBC Motion Gallery / Getty Images

Page 140: Invitation to tea party; from the collection of the Royal Institution (RI MS AD 06/A/03/C/1991/ folder 8)

Page 141: Annotated page of Dawkins's Lecture 1 script; Royal Institution / Richard Dawkins (RI MS AD 06/A/03/C/1991/folder 11)

Page 143: Still from Lecture 1; BBC Motion Gallery / Getty Images

Page 146: Still from Lecture 2; BBC Motion Gallery / Getty Images

Page 154: Still from Lecture 4; BBC Motion Gallery / Getty Images

Page 158: Loan agreement form from NHM; from the collection of the Royal Institution (RI MS AD 06/A/03/C/1996)

Page 161: Still from Lecture 2; Channel 4

Page 163: Still from Lecture 2; Channel 4

Page 166: Still from Lecture 3; Channel 4

Page 169: Hot water drill ice core unit; Photographer Alex Taylor, British Antarctic Survey

Page 173: Still from Lecture 3; Channel 4

Page 178: Photo of giant caterpillar and holly leaf; from the collection of the Royal Institution, photographer unknown

Page 180: Photo of volunteer Adam; from the collection of the Royal Institution, photographer unknown

Page 183: Photo of Hartley during demonstration; from the collection of the Royal Institution, photographer unknown

Page 185: Still from Lecture 3; Furnace TV

Page 187: Still from Lecture 5; Furnace TV

Page 193: Photo of Hartley with donkeys from Lecture series; from the collection of the Royal Institution, photographer unknown

Index

(page numbers in italic type indicate illustrations)

Hoc variare, decus mu
Artificis.

Omne quod Æternus per
Autoris nomen celebret,

.1.